FREE Test Taking Tips DVD Offer

To help us better serve you, we have developed a Test Taking Tips DVD that we would like to give you for FREE. **This DVD covers world-class test taking tips that you can use to be even more successful when you are taking your test.**

All that we ask is that you email us your feedback about your study guide. Please let us know what you thought about it – whether that is good, bad or indifferent.

To get your **FREE Test Taking Tips DVD**, email freedvd@studyguideteam.com with "FREE DVD" in the subject line and the following information in the body of the email:

a. The title of your study guide.

b. Your product rating on a scale of 1-5, with 5 being the highest rating.

c. Your feedback about the study guide. What did you think of it?

d. Your full name and shipping address to send your free DVD.

If you have any questions or concerns, please don't hesitate to contact us at freedvd@studyguideteam.com.

Thanks again!

HAPS Exam Prep

Human Anatomy & Physiology Study Guide with Practice Test Questions

[2nd Edition]

Joshua Rueda

Written and edited by TPB Publishing.

ISBN 13: 9781637751848
ISBN 10: 1637751842

Table of Contents

Quick Overview

As you draw closer to taking your exam, effective preparation becomes more and more important. Thankfully, you have this study guide to help you get ready. Use this guide to help keep your studying on track and refer to it often.

This study guide contains several key sections that will help you be successful on your exam. The guide contains tips for what you should do the night before and the day of the test. Also included are test-taking tips. Knowing the right information is not always enough. Many well-prepared test takers struggle with exams. These tips will help equip you to accurately read, assess, and answer test questions.

A large part of the guide is devoted to showing you what content to expect on the exam and to helping you better understand that content. In this guide are practice test questions so that you can see how well you have grasped the content. Then, answer explanations are provided so that you can understand why you missed certain questions.

Don't try to cram the night before you take your exam. This is not a wise strategy for a few reasons. First, your retention of the information will be low. Your time would be better used by reviewing information you already know rather than trying to learn a lot of new information. Second, you will likely become stressed as you try to gain a large amount of knowledge in a short amount of time. Third, you will be depriving yourself of sleep. So be sure to go to bed at a reasonable time the night before. Being well-rested helps you focus and remain calm.

Be sure to eat a substantial breakfast the morning of the exam. If you are taking the exam in the afternoon, be sure to have a good lunch as well. Being hungry is distracting and can make it difficult to focus. You have hopefully spent lots of time preparing for the exam. Don't let an empty stomach get in the way of success!

When travelling to the testing center, leave earlier than needed. That way, you have a buffer in case you experience any delays. This will help you remain calm and will keep you from missing your appointment time at the testing center.

Be sure to pace yourself during the exam. Don't try to rush through the exam. There is no need to risk performing poorly on the exam just so you can leave the testing center early. Allow yourself to use all of the allotted time if needed.

Remain positive while taking the exam even if you feel like you are performing poorly. Thinking about the content you should have mastered will not help you perform better on the exam.

Once the exam is complete, take some time to relax. Even if you feel that you need to take the exam again, you will be well served by some down time before you begin studying again. It's often easier to convince yourself to study if you know that it will come with a reward!

Test-Taking Strategies

1. Predicting the Answer

When you feel confident in your preparation for a multiple-choice test, try predicting the answer before reading the answer choices. This is especially useful on questions that test objective factual knowledge. By predicting the answer before reading the available choices, you eliminate the possibility that you will be distracted or led astray by an incorrect answer choice. You will feel more confident in your selection if you read the question, predict the answer, and then find your prediction among the answer choices. After using this strategy, be sure to still read all of the answer choices carefully and completely. If you feel unprepared, you should not attempt to predict the answers. This would be a waste of time and an opportunity for your mind to wander in the wrong direction.

2. Reading the Whole Question

Too often, test takers scan a multiple-choice question, recognize a few familiar words, and immediately jump to the answer choices. Test authors are aware of this common impatience, and they will sometimes prey upon it. For instance, a test author might subtly turn the question into a negative, or he or she might redirect the focus of the question right at the end. The only way to avoid falling into these traps is to read the entirety of the question carefully before reading the answer choices.

3. Looking for Wrong Answers

Long and complicated multiple-choice questions can be intimidating. One way to simplify a difficult multiple-choice question is to eliminate all of the answer choices that are clearly wrong. In most sets of answers, there will be at least one selection that can be dismissed right away. If the test is administered on paper, the test taker could draw a line through it to indicate that it may be ignored; otherwise, the test taker will have to perform this operation mentally or on scratch paper. In either case, once the obviously incorrect answers have been eliminated, the remaining choices may be considered. Sometimes identifying the clearly wrong answers will give the test taker some information about the correct answer. For instance, if one of the remaining answer choices is a direct opposite of one of the eliminated answer choices, it may well be the correct answer. The opposite of obviously wrong is obviously right! Of course, this is not always the case. Some answers are obviously incorrect simply because they are irrelevant to the question being asked. Still, identifying and eliminating some incorrect answer choices is a good way to simplify a multiple-choice question.

4. Don't Overanalyze

Anxious test takers often overanalyze questions. When you are nervous, your brain will often run wild, causing you to make associations and discover clues that don't actually exist. If you feel that this may be a problem for you, do whatever you can to slow down during the test. Try taking a deep breath or counting to ten. As you read and consider the question, restrict yourself to the particular words used by the author. Avoid thought tangents about what the author *really* meant, or what he or she was *trying* to say. The only things that matter on a multiple-choice test are the words that are actually in the question. You must avoid reading too much into a multiple-choice question, or supposing that the writer meant something other than what he or she wrote.

5. No Need for Panic

It is wise to learn as many strategies as possible before taking a multiple-choice test, but it is likely that you will come across a few questions for which you simply don't know the answer. In this situation, avoid panicking. Because most multiple-choice tests include dozens of questions, the relative value of a single wrong answer is small. As much as possible, you should compartmentalize each question on a multiple-choice test. In other words, you should not allow your feelings about one question to affect your success on the others. When you find a question that you either don't understand or don't know how to answer, just take a deep breath and do your best. Read the entire question slowly and carefully. Try rephrasing the question a couple of different ways. Then, read all of the answer choices carefully. After eliminating obviously wrong answers, make a selection and move on to the next question.

6. Confusing Answer Choices

When working on a difficult multiple-choice question, there may be a tendency to focus on the answer choices that are the easiest to understand. Many people, whether consciously or not, gravitate to the answer choices that require the least concentration, knowledge, and memory. This is a mistake. When you come across an answer choice that is confusing, you should give it extra attention. A question might be confusing because you do not know the subject matter to which it refers. If this is the case, don't eliminate the answer before you have affirmatively settled on another. When you come across an answer choice of this type, set it aside as you look at the remaining choices. If you can confidently assert that one of the other choices is correct, you can leave the confusing answer aside. Otherwise, you will need to take a moment to try to better understand the confusing answer choice. Rephrasing is one way to tease out the sense of a confusing answer choice.

7. Your First Instinct

Many people struggle with multiple-choice tests because they overthink the questions. If you have studied sufficiently for the test, you should be prepared to trust your first instinct once you have carefully and completely read the question and all of the answer choices. There is a great deal of research suggesting that the mind can come to the correct conclusion very quickly once it has obtained all of the relevant information. At times, it may seem to you as if your intuition is working faster even than your reasoning mind. This may in fact be true. The knowledge you obtain while studying may be retrieved from your subconscious before you have a chance to work out the associations that support it. Verify your instinct by working out the reasons that it should be trusted.

8. Key Words

Many test takers struggle with multiple-choice questions because they have poor reading comprehension skills. Quickly reading and understanding a multiple-choice question requires a mixture of skill and experience. To help with this, try jotting down a few key words and phrases on a piece of scrap paper. Doing this concentrates the process of reading and forces the mind to weigh the relative importance of the question's parts. In selecting words and phrases to write down, the test taker thinks about the question more deeply and carefully. This is especially true for multiple-choice questions that are preceded by a long prompt.

9. Subtle Negatives

One of the oldest tricks in the multiple-choice test writer's book is to subtly reverse the meaning of a question with a word like *not* or *except*. If you are not paying attention to each word in the question, you can easily be led astray by this trick. For instance, a common question format is, "Which of the following is...?" Obviously, if the question instead is, "Which of the following is not...?," then the answer will be quite different. Even worse, the test makers are aware of the potential for this mistake and will include one answer choice that would be correct if the question were not negated or reversed. A test taker who misses the reversal will find what he or she believes to be a correct answer and will be so confident that he or she will fail to reread the question and discover the original error. The only way to avoid this is to practice a wide variety of multiple-choice questions and to pay close attention to each and every word.

10. Reading Every Answer Choice

It may seem obvious, but you should always read every one of the answer choices! Too many test takers fall into the habit of scanning the question and assuming that they understand the question because they recognize a few key words. From there, they pick the first answer choice that answers the question they believe they have read. Test takers who read all of the answer choices might discover that one of the latter answer choices is actually *more* correct. Moreover, reading all of the answer choices can remind you of facts related to the question that can help you arrive at the correct answer. Sometimes, a misstatement or incorrect detail in one of the latter answer choices will trigger your memory of the subject and will enable you to find the right answer. Failing to read all of the answer choices is like not reading all of the items on a restaurant menu: you might miss out on the perfect choice.

11. Spot the Hedges

One of the keys to success on multiple-choice tests is paying close attention to every word. This is never truer than with words like almost, most, some, and sometimes. These words are called "hedges" because they indicate that a statement is not totally true or not true in every place and time. An absolute statement will contain no hedges, but in many subjects, the answers are not always straightforward or absolute. There are always exceptions to the rules in these subjects. For this reason, you should favor those multiple-choice questions that contain hedging language. The presence of qualifying words indicates that the author is taking special care with their words, which is certainly important when composing the right answer. After all, there are many ways to be wrong, but there is only one way to be right! For this reason, it is wise to avoid answers that are absolute when taking a multiple-choice test. An absolute answer is one that says things are either all one way or all another. They often include words like *every*, *always*, *best*, and *never*. If you are taking a multiple-choice test in a subject that doesn't lend itself to absolute answers, be on your guard if you see any of these words.

12. Long Answers

In many subject areas, the answers are not simple. As already mentioned, the right answer often requires hedges. Another common feature of the answers to a complex or subjective question are qualifying clauses, which are groups of words that subtly modify the meaning of the sentence. If the question or answer choice describes a rule to which there are exceptions or the subject matter is complicated, ambiguous, or confusing, the correct answer will require many words in order to be expressed clearly and accurately. In essence, you should not be deterred by answer choices that seem excessively long. Oftentimes, the author of the text will not be able to write the correct answer without offering some qualifications and modifications. Your job is to read the answer choices thoroughly and completely and to select the one that most accurately and precisely answers the question.

13. Restating to Understand

Sometimes, a question on a multiple-choice test is difficult not because of what it asks but because of how it is written. If this is the case, restate the question or answer choice in different words. This process serves a couple of important purposes. First, it forces you to concentrate on the core of the question. In order to rephrase the question accurately, you have to understand it well. Rephrasing the question will concentrate your mind on the key words and ideas. Second, it will present the information to your mind in a fresh way. This process may trigger your memory and render some useful scrap of information picked up while studying.

14. True Statements

Sometimes an answer choice will be true in itself, but it does not answer the question. This is one of the main reasons why it is essential to read the question carefully and completely before proceeding to the answer choices. Too often, test takers skip ahead to the answer choices and look for true statements. Having found one of these, they are content to select it without reference to the question above. Obviously, this provides an easy way for test makers to play tricks. The savvy test taker will always read the entire question before turning to the answer choices. Then, having settled on a correct answer choice, he or she will refer to the original question and ensure that the selected answer is relevant. The mistake of choosing a correct-but-irrelevant answer choice is especially common on questions related to specific pieces of objective knowledge. A prepared test taker will have a wealth of factual knowledge at their disposal, and should not be careless in its application.

15. No Patterns

One of the more dangerous ideas that circulates about multiple-choice tests is that the correct answers tend to fall into patterns. These erroneous ideas range from a belief that B and C are the most common right answers, to the idea that an unprepared test-taker should answer "A-B-A-C-A-D-A-B-A." It cannot be emphasized enough that pattern-seeking of this type is exactly the WRONG way to approach a multiple-choice test. To begin with, it is highly unlikely that the test maker will plot the correct answers according to some predetermined pattern. The questions are scrambled and delivered in a random order. Furthermore, even if the test maker was following a pattern in the assignation of correct answers, there is no reason why the test taker would know which pattern he or she was using. Any attempt to discern a pattern in the answer choices is a waste of time and a distraction from the real work of taking the test. A test taker would be much better served by extra preparation before the test than by reliance on a pattern in the answers.

FREE DVD OFFER

Don't forget that doing well on your exam includes both understanding the test content and understanding how to use what you know to do well on the test. We offer a completely FREE Test Taking Tips DVD that covers world class test taking tips that you can use to be even more successful when you are taking your test.

All that we ask is that you email us your feedback about your study guide. To get your **FREE Test Taking Tips DVD**, email freedvd@studyguideteam.com with "FREE DVD" in the subject line and the following information in the body of the email:

- The title of your study guide.
- Your product rating on a scale of 1-5, with 5 being the highest rating.
- Your feedback about the study guide. What did you think of it?
- Your full name and shipping address to send your free DVD.

Body Plan and Organization

General Anatomy and Physiology of a Human

Anatomy may be defined as the structural makeup of an organism. The study of anatomy may be divided into microscopic/fine anatomy and macroscopic/gross anatomy. Fine anatomy concerns itself with viewing the features of the body with the aid of a microscope, while gross anatomy concerns itself with viewing the features of the body with the naked eye. Physiology refers to the functions of an organism, and it examines the chemical or physical functions that help the body function appropriately.

Levels of Organization of the Human Body

All the parts of the human body are built of individual units called *cells*. Groups of similar cells are arranged into *tissues,* different tissues are arranged into *organs,* and organs working together form entire *organ systems.* The human body has twelve organ systems that govern circulation, digestion, immunity, hormones, movement, support, coordination, urination & excretion, reproduction (male and female), respiration, and general protection.

Here are some of the systems of the body:

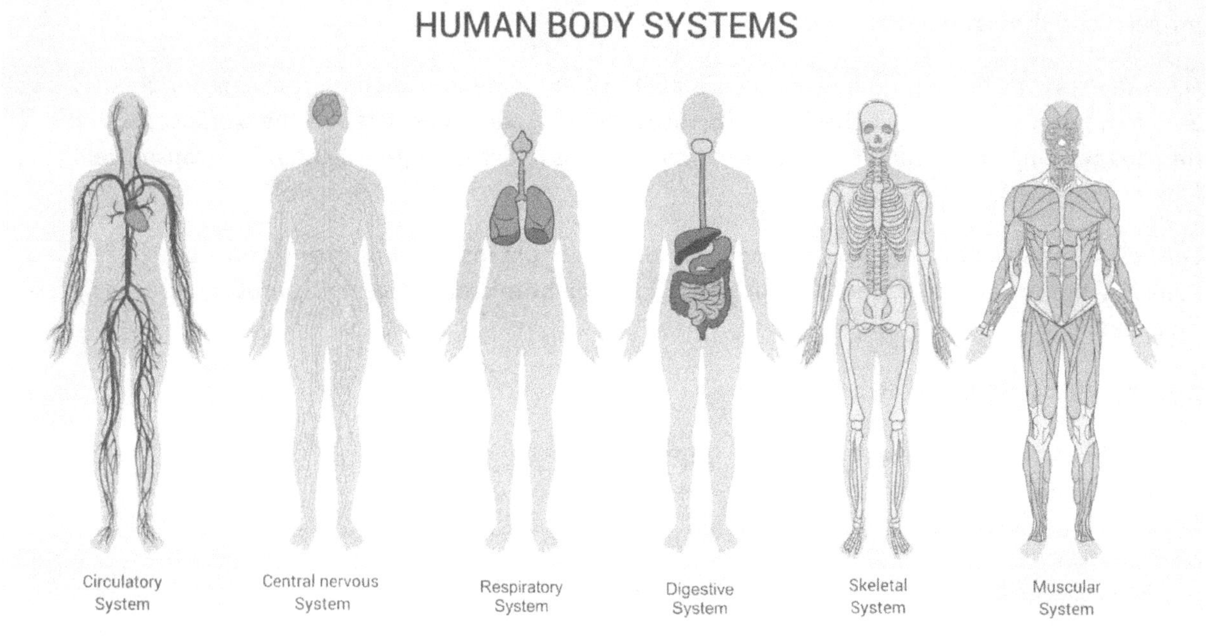
HUMAN BODY SYSTEMS
Circulatory System
Central nervous System
Respiratory System
Digestive System
Skeletal System
Muscular System

Body Cavities

The body is partitioned into different hollow spaces that house organs. The human body contains the following cavities:

- Cranial cavity: The cranial cavity is surrounded by the skull and contains organs such as the brain and pituitary gland.
- Thoracic cavity: The thoracic cavity is encircled by the sternum (breastbone) and ribs. It contains organs such as the lungs, heart, trachea (windpipe), esophagus, and bronchial tubes.

- Abdominal cavity: The abdominal cavity is separated from the thoracic cavity by the diaphragm. It contains organs such as the stomach, gallbladder, liver, small intestines, and large intestines. The abdominal organs are held in place by a membrane called the peritoneum.

- Pelvic cavity: The pelvic cavity is enclosed by the pelvis, or bones of the hip. It contains organs such as the urinary bladder, urethra, ureters, anus, and rectum. It contains the reproductive organs as well. In females, the pelvic cavity also contains the uterus.

- Spinal cavity: The spinal cavity is surrounded by the vertebral column. The vertebral column has five regions: cervical, thoracic, lumbar, sacral, and coccygeal. The spinal cord runs through the middle of the spinal cavity.

Human Tissues

Human tissues can be grouped into four categories:

Muscle: Muscle tissue supports the body and allows it to move, and muscles are special because their cells have the ability to contract. There are three distinct types of muscle tissue: skeletal, smooth, and cardiac. Skeletal muscle is voluntary, or under conscious control, and is usually attached to bones. Most body movement is directly caused by the contraction of skeletal muscle. Smooth muscle is typically involuntary, or not under conscious control, and may be found in blood vessels, the walls of hollow organs, and the urinary bladder. Cardiac muscle is involuntary and found in the heart, which helps propel blood throughout the body.

Nervous: Nervous tissue is unique in that it is able to coordinate information from sensory organs as well as communicate the proper behavioral responses. Neurons, or nerve cells, are the workhorses of the nervous system. They communicate via action potentials (electrical signals) and neurotransmitters (chemical signals).

Epithelial: Epithelial tissue covers the external surfaces of organs and lines many of the body's cavities. Epithelial tissue helps to protect the body from invasion by microbes (bacteria, viruses, parasites), fluid loss, and injury.

Epithelial cell shapes can be:

- Squamous: cells with a flat shape
- Cuboidal: cells with a cubed shape
- Columnar: cells shaped like a column

Epithelial cells can be arranged in four patterns:

- Simple: a type of epithelium composed solely from a single layer of cells
- Stratified: a type of epithelium composed of multiple layers of cells
- Pseudostratified: a type of epithelium that appears to be stratified but actually consists of only one layer of cells
- Transitional: a type of epithelium noted for its ability to expand and contract

Connective: Connective tissue supports and connects the tissues and organs of the body. Connective tissue is composed of cells dispersed throughout a matrix which can be gel, liquid, protein fibers, or salts. The primary protein fibers in the matrix are collagen (for strength), elastin (for flexibility), and reticulum (for support). Connective tissue can be categorized as either *loose* or *dense*. Examples of connective tissue include bones, cartilage, ligaments, tendons, blood, and adipose (fat) tissue.

Three Primary Body Planes

A plane is an imaginary flat surface. The three primary planes of the human body are frontal, sagittal, and transverse. The coronal plane is a vertical plane that divides the body or organ into front (anterior) and back (posterior) portions. The **sagittal plane**, or lateral plane, is a vertical plane that divides the body or organ into right and left sides. The transverse plane is a horizontal plane that divides the body or organ into upper and lower portions. In medical imaging, computed tomography (CT) scans are oriented only in the transverse plane; while magnetic resonance imaging (MRI) scans may be oriented in any of the three planes.

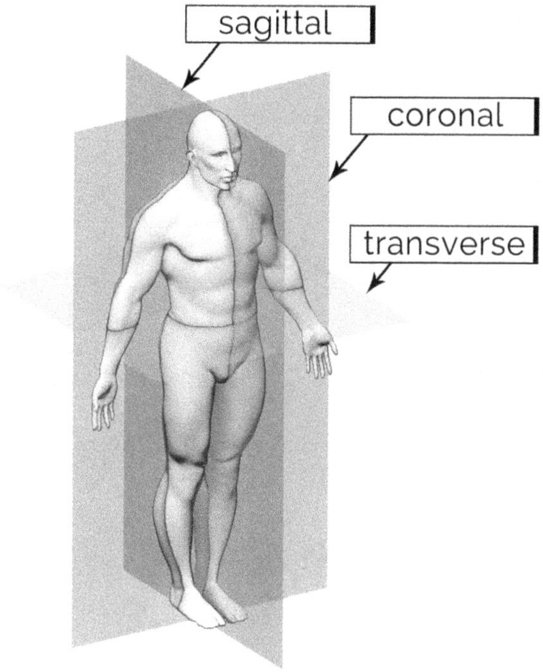

Terms of Direction

Medial refers to a structure being closer to the midline of the body. For example, the nose is medial to the eyes.

Lateral refers to a structure being farther from the midline of the body, and it is the opposite of *medial*. For example, the eyes are lateral to the nose.

Proximal refers to a structure or body part located near an attachment point. For example, the elbow is proximal to the wrist.

Distal refers to a structure or body part located far from an attachment point, and it is the opposite of *proximal*. For example, the wrist is distal to the elbow.

Anterior means toward the front in humans. For example, the lips are anterior to the teeth. The term *ventral* can be used in place of *anterior*.

Posterior means toward the back in humans, and it is the opposite of *anterior*. For example, the teeth are posterior to the lips. The term *dorsal* can be used in place of *posterior*.

Superior means above and refers to a structure closer to the head. For example, the head is superior to the neck. The terms *cephalic* or *cranial* may be used in place of *superior*.

Inferior means below and refers to a structure farther from the head, and it is the opposite of *superior*. For example, the neck is inferior to the head. The term *caudal* may be used in place of *inferior*.

Superficial refers to a structure closer to the surface. For example, the muscles are superficial because they are just beneath the surface of the skin.

Deep refers to a structure farther from the surface, and it is the opposite of *superficial*. For example, the femur is a deep structure lying beneath the muscles.

Body Regions

- Terms for general locations on the body include:
- Cervical: relating to the neck
- Clavicular: relating to the clavicle, or collarbone
- Ocular: relating to the eyes
- Acromial: relating to the shoulder
- Cubital: relating to the elbow
- Brachial: relating to the arm
- Carpal: relating to the wrist
- Thoracic : relating to the chest
- Abdominal: relating to the abdomen
- Pubic: relating to the groin
- Pelvic: relating to the pelvis, or bones of the hip
- Femoral: relating to the femur, or thigh bone
- Geniculate: relating to the knee
- Pedal: relating to the foot
- Palmar: relating to the palm of the hand
- Plantar: relating to the sole of the foot

Abdominopelvic Regions and Quadrants

The abdominopelvic region may be defined as the combination of the abdominal and the pelvic cavities. The region's upper border is the breasts and its lower border is the groin region. The region is divided into the following nine sections:

- Right hypochondriac: region below the cartilage of the ribs
- Epigastric: region above the stomach between the hypochondriac regions
- Left hypochondriac: region below the cartilage of the ribs
- Right lumbar: region of the waist
- Umbilical: region between the lumbar regions where the umbilicus, or belly button (navel), is located
- Left lumbar: region of the waist
- Right inguinal: region of the groin

- Hypogastric: region below the stomach between the inguinal regions
- Left inguinal: region of the groin

A simpler way to describe the abdominopelvic area is to divide it into the following quadrants:

- Right upper quadrant (RUQ): Encompasses the right hypochondriac, right lumbar, epigastric, and umbilical regions.
- Right lower quadrant (RLQ): Encompasses the right lumbar, right inguinal, hypogastric, and umbilical regions.
- Left upper quadrant (LUQ): Encompasses the left hypochondriac, left lumbar, epigastric, and umbilical regions.
- Left lower quadrant (LLQ): Encompasses the left lumbar, left inguinal, hypogastric, and umbilical regions.

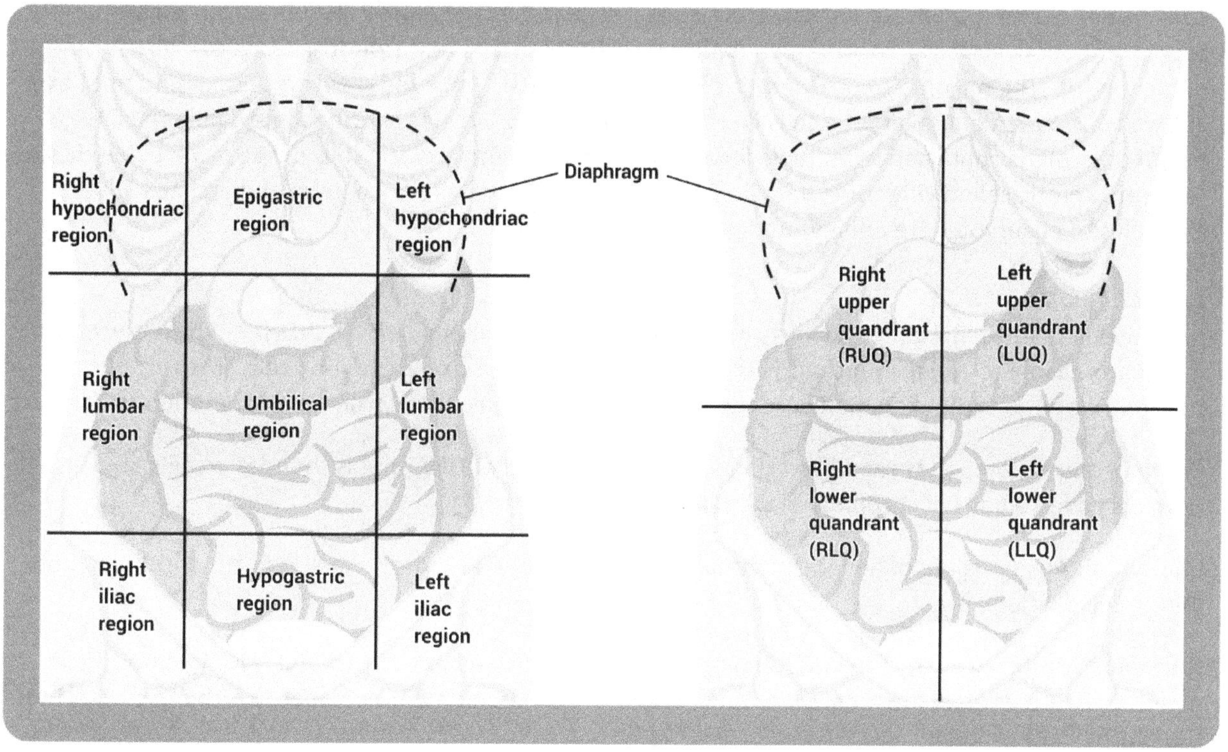

Cell Structure and Function

The cell is the main functional and structural component of all living organisms. Robert Hooke, an English scientist, coined the term "cell" in 1665. Hooke's discovery laid the groundwork for the cell theory, which is composed of three principals:

1. All organisms are composed of cells.

2. All existing cells are created from other living cells.

3. The cell is the most fundamental unit of life.

Organisms can be unicellular (composed of one cell) or multicellular (composed of many cells). All cells must be bounded by a cell membrane, be filled with cytoplasm of some sort, and be coded by a genetic sequence.

The cell membrane separates a cell's internal and external environments. It is a selectively permeable membrane, which usually only allows the passage of certain molecules by diffusion. Phospholipids and proteins are crucial components of all cell membranes. The cytoplasm is the cell's internal environment and is aqueous, or water-based. The genome represents the genetic material inside the cell that is passed on from generation to generation.

Prokaryotes and Eukaryotes

Prokaryotic cells are much smaller than eukaryotic cells. The majority of prokaryotes are unicellular, while the majority of eukaryotes are multicellular. Prokaryotic cells have no nucleus, and their genome is found in an area known as the nucleoid. They also do not have membrane-bound organelles, which are "little organs" that perform specific functions within a cell.

Eukaryotic cells have a proper nucleus containing the genome. They also have numerous membrane-bound organelles such as lysosomes, endoplasmic reticula (rough and smooth), Golgi complexes, and mitochondria. The majority of prokaryotic cells have cell walls, while most eukaryotic cells do not have cell walls. The DNA of prokaryotic cells is contained in a single circular chromosome, while the DNA of eukaryotic cells is contained in multiple linear chromosomes. Prokaryotic cells divide using binary fission, while eukaryotic cells divide using mitosis. Examples of prokaryotes are bacteria and archaea while examples of eukaryotes are animals and plants.

Nuclear Parts of a Cell

Nucleus (plural nuclei): Houses a cell's genetic material, deoxyribonucleic acid (DNA), which is used to form chromosomes. A single nucleus is the defining characteristic of eukaryotic cells. The nucleus of a cell controls gene expression. It ensures genetic material is transmitted from one generation to the next.

Chromosomes: Complex thread-like arrangements composed of DNA that is found in a cell's nucleus. Humans have 23 pairs of chromosomes for a total of 46.

Chromatin: An aggregate of genetic material consisting of DNA and proteins that forms chromosomes during cell division.

Nucleolus (plural nucleoli): The largest component of the nucleus of a eukaryotic cell. With no membrane, the primary function of the nucleolus is the production of ribosomes, which are crucial to the synthesis of proteins.

Cell Membranes

Cell membranes encircle the cell's cytoplasm, separating the intracellular environment from the extracellular environment. They are selectively permeable, which enables them to control molecular traffic entering and exiting cells. Cell membranes are made of a double layer of phospholipids studded with proteins. Cholesterol is also dispersed in the phospholipid bilayer of cell membranes to provide stability. The proteins in the phospholipid bilayer aid the transport of molecules across cell membranes.

Scientists use the term "fluid mosaic model" to refer to the arrangement of phospholipids and proteins in cell membranes. In that model, phospholipids have a head region and a tail region. The head region of the phospholipids is attracted to water (hydrophilic), while the tail region is repelled by it (hydrophobic).

Because they are hydrophilic, the heads of the phospholipids are facing the water, pointing inside and outside of the cell. Because they are hydrophobic, the tails of the phospholipids are oriented inward between both head regions. This orientation constructs the phospholipid bilayer.

Cell membranes have the distinct trait of selective permeability. The fact that cell membranes are amphiphilic (having hydrophilic and hydrophobic zones) contributes to this trait. As a result, cell membranes are able to regulate the flow of molecules in and out of the cell.

Factors relating to molecules such as size, polarity, and solubility determine their likelihood of passage across cell membranes. Small molecules are able to diffuse easily across cell membranes compared to large molecules. Polarity refers to the charge present in a molecule. Polar molecules have regions, or poles, of positive and negative charge and are water soluble, while nonpolar molecules have no charge and are fat-soluble. Solubility refers to the ability of a substance, called a solute, to dissolve in a solvent. A soluble substance can be dissolved in a solvent, while an insoluble substance cannot be dissolved in a solvent. Nonpolar, fat-soluble substances have a much easier time passing through cell membranes compared to polar, water-soluble substances.

Passive Transport Mechanisms

Passive transport refers to the migration of molecules across a cell membrane that does not require energy. The three types of passive transport include simple diffusion, facilitated diffusion, and osmosis.

Simple diffusion relies on a concentration gradient, or differing quantities of molecules inside or outside of a cell. During simple diffusion, molecules move from an area of high concentration to an area of low concentration. Facilitated diffusion utilizes carrier proteins to transport molecules across a cell membrane. Osmosis refers to the transport of water across a selectively permeable membrane. During osmosis, water moves from a region of low solute concentration to a region of high solute concentration.

Active Transport Mechanisms

Active transport refers to the energy-requiring migration of molecules across a cell membrane. It's a useful way to move molecules from an area of low concentration to an area of high concentration. Adenosine triphosphate (ATP), the currency of cellular energy, is needed to work against the concentration gradient.

Active transport can involve carrier proteins that cross the cell membrane to pump molecules and ions across the membrane, like in facilitated diffusion. The difference is that active transport uses the energy from ATP to drive this transport, as typically the ions or molecules are going against their concentration gradient. For example, glucose pumps in the kidney pump all of the glucose into the cells from the lumen of the nephron even though there is a higher concentration of glucose in the cell than in the lumen. This is because glucose is a precious fuel source, and the body wants to conserve as much as possible. Pumps can either send a molecule in one direction, multiple molecules in the same direction (symports), or multiple molecules in different directions (antiports).

Active transport can also involve the movement of membrane-bound particles, either into a cell (endocytosis) or out of a cell (exocytosis). The three major forms of endocytosis are: pinocytosis, where the cell is *drinking* and intakes only small molecules; phagocytosis, where the cell is *eating* and intakes large particles or small organisms; and receptor-mediated endocytosis, where the cell's membrane splits off to form an internal vesicle as a response to molecules activating receptors on its surface. Exocytosis is the inverse of endocytosis, and the membranes of the vesicle join to that of the cell's surface while the molecules inside the vesicle are released outside. This is common in nervous and muscle tissue for the release of neurotransmitters and in endocrine cells for the release of hormones. The two major categories of exocytosis are excretion and secretion. Excretion is defined as the removal of waste from a cell. Secretion is defined as the transport of molecules, such as hormones or enzymes, from a cell.

Structure and Function of Cellular Organelles

Organelles are specialized structures that perform specific tasks in a cell. The term literally means "little organ." Most organelles are membrane bound and serve as sites for the production or degradation of chemicals. The following are organelles found in eukaryotic cells:

Nucleus: A cell's nucleus contains genetic information in the form of DNA. The nucleus is surrounded by the nuclear envelope. A single nucleus is the defining characteristic of eukaryotic cells. The nucleus is also the most important organelle of the cell. It contains the nucleolus, which manufactures ribosomes (another organelle) that are crucial in protein synthesis (also called gene expression).

Mitochondria: Mitochondria are oval-shaped and have a double membrane. The inner membrane has multiple folds called cristae. Mitochondria are responsible for the production of a cell's energy in the form of adenosine triphosphate (ATP). ATP is the principal energy transfer molecule in eukaryotic cells. Mitochondria also participate in cellular respiration.

Rough Endoplasmic Reticulum: The rough endoplasmic reticulum (RER) is composed of linked membranous sacs called cisternae with ribosomes attached to their external surfaces. The RER is responsible for the production of proteins that will eventually get shipped out of the cell.

Smooth Endoplasmic Reticulum: The smooth endoplasmic reticulum (SER) is composed of linked membranous sacs called cisternae without ribosomes, which distinguishes it from the RER. The SER's main function is the production of carbohydrates and lipids, which can be created expressly for the cell, or to modify the proteins from the RER that will eventually get shipped out of the cell.

Golgi Apparatus: The Golgi apparatus is located next to the SER. Its main function is the final modification, storage, and shipping of products (proteins, carbohydrates, and lipids) from the endoplasmic reticulum.

Lysosomes: Lysosomes are specialized vesicles that contain enzymes capable of digesting food, surplus organelles, and foreign invaders such as bacteria and viruses. They often destroy dead cells in order to recycle cellular components. Lysosomes are found only in animal cells.

Secretory Vesicles: Secretory vesicles transport and deliver molecules into or out of the cell via the cell membrane. Endocytosis refers to the movement of molecules into a cell via secretory vesicles. Exocytosis refers to the movement of molecules out of a cell via secretory vesicles.

Ribosomes: Ribosomes are not membrane bound. They are responsible for the production of proteins as specified from DNA instructions. Ribosomes can be free or bound.

Cilia and Flagella: Cilia are specialized hair-like projections on some eukaryotic cells that aid in movement, while flagella are long, whip-like projections that are used in the same capacity.

Here is an illustration of the cell:

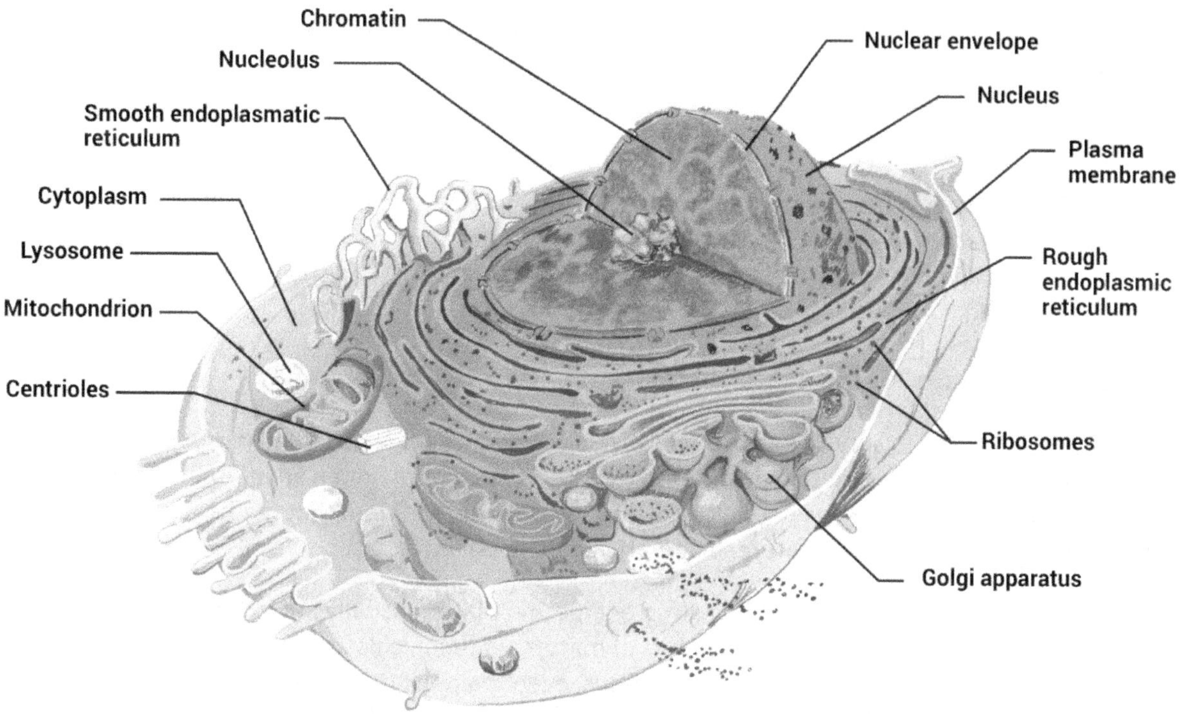

The following organelles are not found in animal cells:

Cell Walls: Cell walls can be found in plants, bacteria, and fungi, and are made of cellulose, peptidoglycan, and lignin, depending on the organism it surrounds. Each of these materials is a type of sugar recognized as a structural carbohydrate. The carbohydrates form rigid structures located outside of the cell membrane. Cell walls protect the cell, help maintain a cell's shape, and provide structural support.

Vacuoles: Plant cells have central vacuoles, which are essentially a membrane surrounding a body of water. They may store nutrients or waste products. Since vacuoles are large, they also help to support the structure of plant cells.

Chloroplasts: Chloroplasts are membrane-bound organelles that perform photosynthesis. They contain structural units called thylakoids. Chlorophyll, a green pigment that circulates within the thylakoids, harnesses light energy (sunlight) and helps convert it into chemical energy (glucose).

Homeostasis

Homeostasis is the ability of a system, such as a cell or organism, to regulate itself so that its internal conditions remain stable even when its external environment may be changing. By maintaining homeostasis, systems are able to function normally even in adverse conditions.

Response to Stimuli and Homeostasis

A *stimulus* is a change in the environment, either internal or external, around an organism that is received by a sensory receptor and causes the organism to react. *Homeostasis* is the stable state of an organism. When an organism reacts to stimuli, it works to counteract the change in order to reach homeostasis again.

Entropy and Maintaining Homeostasis

In order for organisms to survive against this universal tendency for chaos, they must use energy to work against entropy. They do this via biochemical processes that maintain an internal order, called homeostasis. Homeostasis is the physiological processes within a system that regulate a stable internal equilibrium such as body temperature, blood pH, and fluid balance.

Organisms maintain homeostasis by using free energy and matter, usually in the form of atoms and molecules. Atoms are the smallest whole units of matter, and molecules are atoms glued to each other via chemical bonds. Everything is made up of atoms, even the most complicated structures; their complexity is only due to the variety and quantity of different molecular arrangements.

Chemistry

Periodicity and States of Matter

Periodic Table of the Elements

Using the periodic table, elements are arranged by atomic number, similar characteristics, and electron configurations in a tabular format. The columns, called *groups*, are sorted by similar chemical properties and characteristics such as appearance and reactivity. This can be seen in the shiny texture of metals, the high melting points of alkali Earth metals, and the softness of post-transition metals. The rows are arranged by electron valance configurations and are called *periods*.

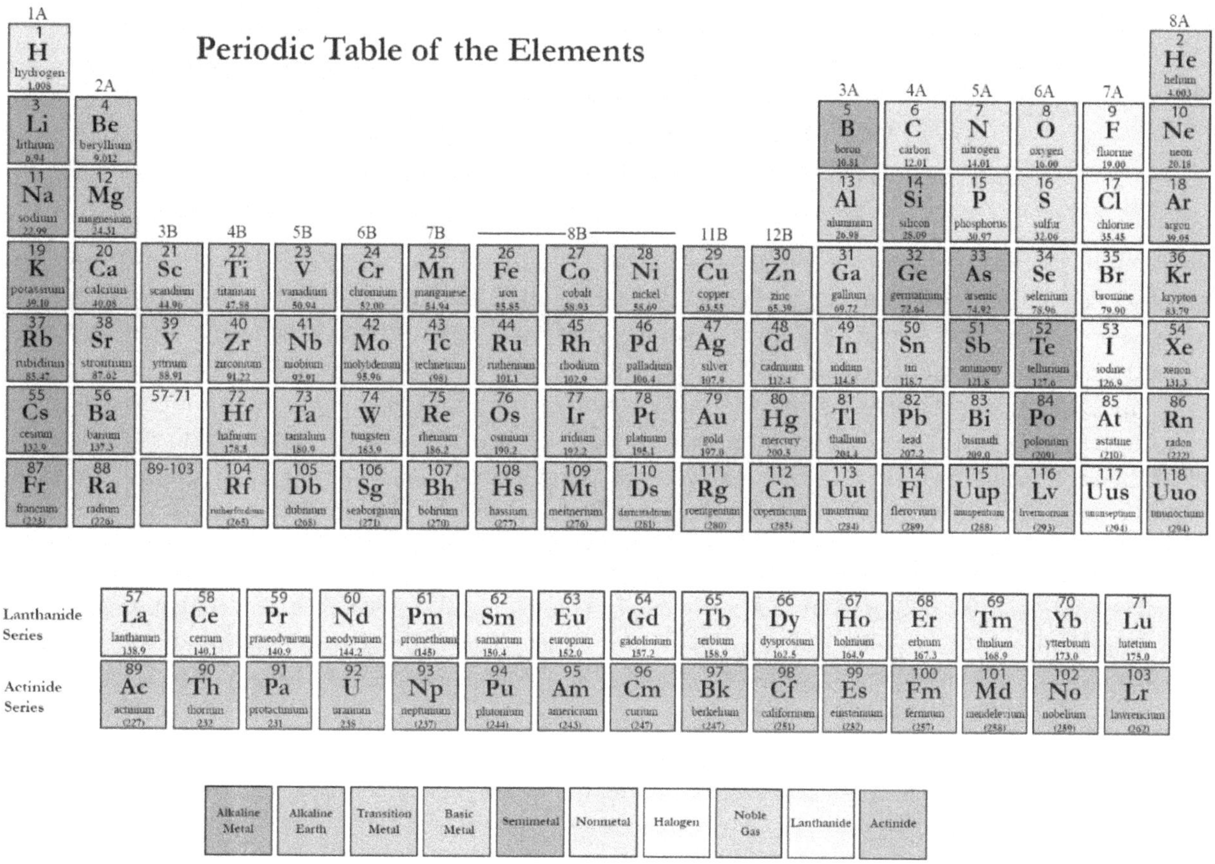

Periodic Table of the Elements
1A 2A 3B 4B 5B 6B 7B 8B 11B 12B 3A 4A 5A 6A 7A 8A
1 H hydrogen
2 He helium
3 Li lithium
4 Be beryllium
5 B boron
6 C carbon
7 N nitrogen
8 O oxygen
9 F fluorine
10 Ne neon
11 Na sodium
12 Mg magnesium
13 Al aluminum
14 Si silicon
15 P phosphorus
16 S sulfur
17 Cl chlorine
18 Ar argon
19 K potassium
20 Ca calcium
21 Sc scandium
22 Ti titanium
23 V vanadium
24 Cr chromium
25 Mn manganese
26 Fe iron
27 Co cobalt
28 Ni nickel
29 Cu copper
30 Zn zinc
31 Ga gallium
32 Ge germanium
33 As arsenic
34 Se selenium
35 Br bromine
36 Kr krypton
37 Rb rubidium
38 Sr strontium
39 Y yttrium
40 Zr zirconium
41 Nb niobium
42 Mo molybdenum
43 Tc technetium
44 Ru ruthenium
45 Rh rhodium
46 Pd palladium
47 Ag silver
48 Cd cadmium
49 In indium
50 Sn tin
51 Sb antimony
52 Te tellurium
53 I iodine
54 Xe xenon
55 Cs cesium
56 Ba barium
57-71
72 Hf hafnium
73 Ta tantalum
74 W tungsten
75 Re rhenium
76 Os osmium
77 Ir iridium
78 Pt platinum
79 Au gold
80 Hg mercury
81 Tl thallium
82 Pb lead
83 Bi bismuth
84 Po polonium
85 At astatine
86 Rn radon
87 Fr francium
88 Ra radium
89-103
104 Rf rutherfordium
105 Db dubnium
106 Sg seaborgium
107 Bh bohrium
108 Hs hassium
109 Mt meitnerium
110 Ds darmstadtium
111 Rg roentgenium
112 Cn copernicium
113 Uut ununtrium
114 Fl flerovium
115 Uup ununpentium
116 Lv livermorium
117 Uus ununseptium
118 Uuo ununoctium
Lanthanide Series
57 La lanthanum
58 Ce cerium
59 Pr praseodymium
60 Nd neodymium
61 Pm promethium
62 Sm samarium
63 Eu europium
64 Gd gadolinium
65 Tb terbium
66 Dy dysprosium
67 Ho holmium
68 Er erbium
69 Tm thulium
70 Yb ytterbium
71 Lu lutetium
Actinide Series
89 Ac actinium
90 Th thorium
91 Pa protactinium
92 U uranium
93 Np neptunium
94 Pu plutonium
95 Am americium
96 Cm curium
97 Bk berkelium
98 Cf californium
99 Es einsteinium
100 Fm fermium
101 Md mendelevium
102 No nobelium
103 Lr lawrencium
Alkaline Metal
Alkaline Earth
Transition Metal
Basic Metal
Semimetal
Nonmetal
Halogen
Noble Gas
Lanthanide
Actinide

The elements are set in ascending order from left to right by atomic number. As mentioned, the atomic number is the number of protons contained within the nucleus of the atom. For example, the element helium has an atomic number of 2 because it has two protons in its nucleus.

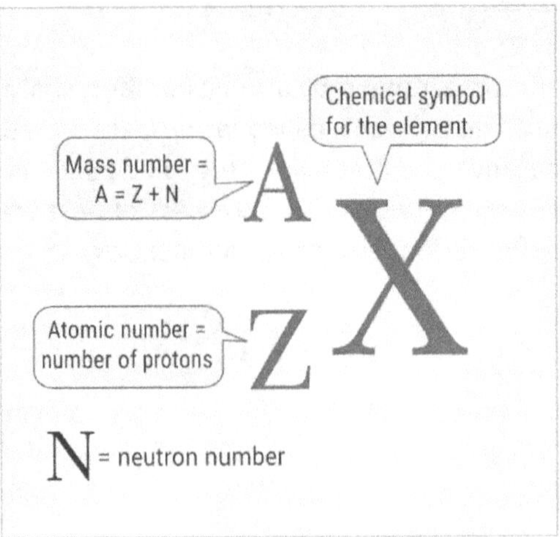

An element's mass number is calculated by adding the number of protons and neutrons of an atom together, while the atomic mass of an element is the weighted average of the naturally occurring atoms of a given element, or the relative abundance of isotopes that might be used in chemistry. For example, the atomic (mass) number of chlorine is 35; however, the atomic mass of chlorine is 35.5 amu (atomic mass unit). This discrepancy exists because there are many isotopes (meaning the nucleus could have 36 instead of 35 protons) occurring in nature. Given the prevalence of the various isotopes, the average of all of the atomic masses turns out to be 35.5 amu, which is slightly higher than chlorine's number on the periodic table. As another example, carbon has an atomic number of 12, but its atomic mass is 12.01 amu because, unlike chlorine, there are few naturally occurring isotopes to raise the average number.

Elements are arranged according to their valance electron configurations, which also contribute to trends in chemical properties. These properties help to further categorize the elements into blocks, including metals, non-metals, transition metals, alkali metals, alkali earth metals, metalloids, lanthanides, actinides, diatomics, post-transition metals, polyatomic non-metals, and noble gases. Noble gases (the far-right column) have a full outer electron valence shell. The elements in this block possess similar characteristics such as being colorless, odorless, and having low chemical reactivity. Another block, the metals, tend to be shiny, highly conductive, and easily form alloys with each other, non-metals, and noble gases.

The symbols of the elements on the periodic table are a single letter or a two-letter combination that is usually derived from the element's name. Many of the elements have Latin origins for their names, and their atomic symbols do not match their modern names. For example, iron is derived from the word *ferrum*, so its symbol is Fe, even though it is now called iron. The naming of the elements began with those of natural origin and their ancient names, which included the use of the ending "ium." This naming practice has been continued for all elements that have been named since the 1940s. Now, the names of new elements must be approved by the International Union of Pure and Applied Chemistry.

The elements on the periodic table are arranged by number and grouped by trends in their physical properties and electron configurations. Certain trends are easily described by the arrangement of the periodic table, which includes the increase of the atomic radius as elements go from right to left and from top to bottom on the periodic table. Another trend on the periodic table is the increase in ionization energy (or the tendency of an atom to attract and form bonds with electrons). This tendency increases from left to right and from bottom to top of the periodic table—the opposite directions of the trend for the atomic radius. The elements on the right side and near the bottom of the periodic table tend to attract electrons with the intent to gain, while the elements on the left and near the top usually lose, or give up, one or more electrons in order to bond. The only exceptions to this rule are the noble gases. Since the noble gases have full valence shells, they do not have a tendency to lose or gain electrons.

Chemical reactivity is another trend identifiable by the groupings of the elements on the periodic table. The chemical reactivity of metals decreases from left to right and while going higher on the table. Conversely, non-metals increase in chemical reactivity from left to right and while going lower on the table. Again, the noble gases present an exception to these trends because they have very low chemical reactivity.

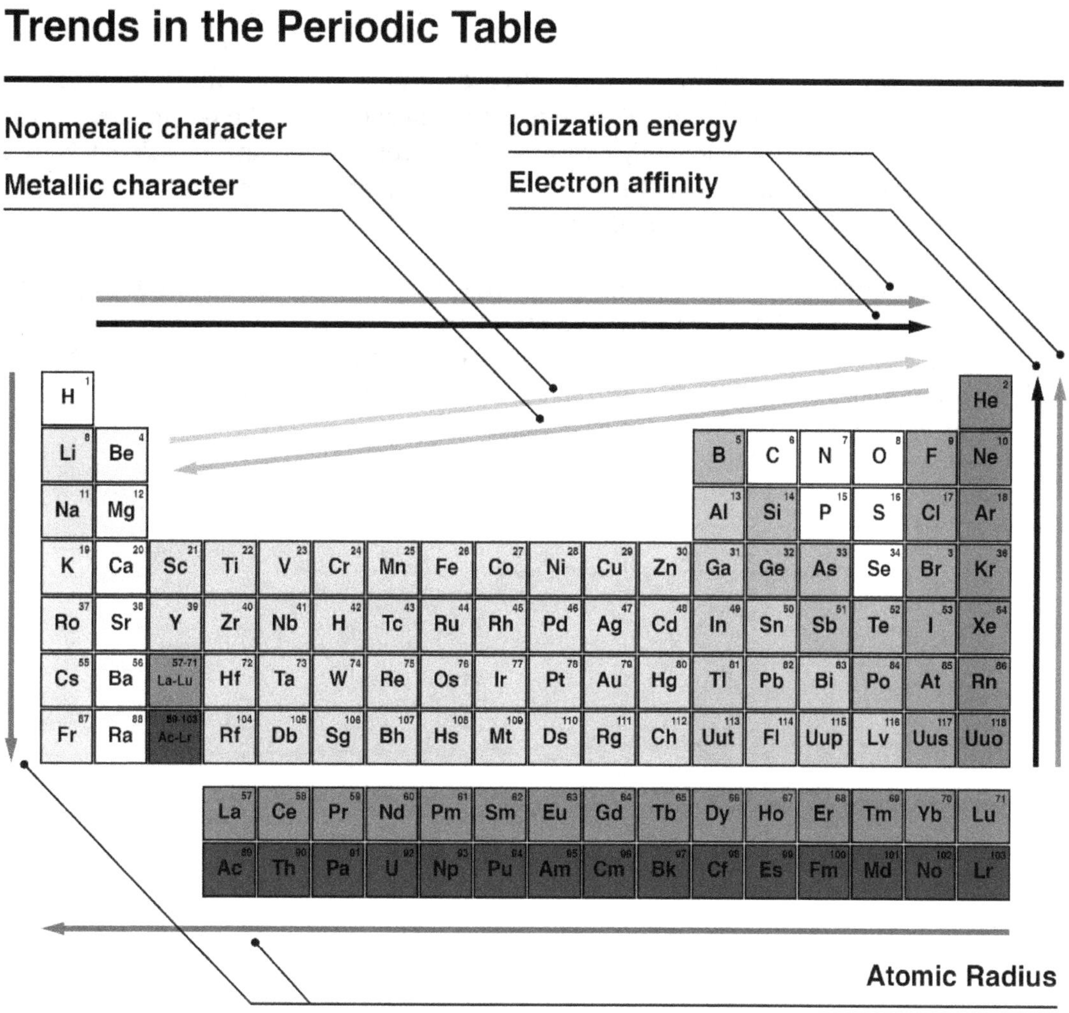

States of Matter and Factors that Affect Phase Changes

Matter is most commonly found in three distinct states: solid, liquid, and gas. A solid has a distinct shape and a defined volume. A liquid has a more loosely defined shape and a definite volume, while a gas has no definite shape or volume. The *Kinetic Theory of Matter* states that matter is composed of a large number of small particles (specifically, atoms and molecules) that are in constant motion. The distance between the separations in these particles determines the state of the matter: solid, liquid, or gas. In gases, the particles have a large separation and no attractive forces. In liquids, there is moderate separation between particles and some attractive forces to form a loose shape. Solids have almost no separation between their particles, causing a defined and set shape. The constant movement of particles causes them to bump into each other, thus allowing the particles to transfer energy between each other. This bumping and transferring of energy helps explain the transfer of heat and the relationship between pressure, volume, and temperature.

The *Ideal Gas Law* states that pressure, volume, and temperature are all related through the equation: $PV = nRT$, where P is pressure, V is volume, n is the amount of the substance in moles, R is the gas constant, and T is temperature.

Through this relationship, volume and pressure are both proportional to temperature, but pressure is inversely proportional to volume. Therefore, if the equation is balanced, and the volume decreases in the system, pressure needs to proportionately increase to keep both sides of the equation balanced. In contrast, if the equation is unbalanced and the pressure increases, then the temperature would also increase, since pressure and temperature are directly proportional.

When pressure, temperature, or volume change in matter, a change in state can occur. Changes in state include solid to liquid (melting), liquid to gas (evaporation), solid to gas (sublimation), gas to solid (deposition), gas to liquid (condensation), and liquid to solid (freezing). There is one other state of matter called *plasma*, which is seen in lightning, television screens, and neon lights. Plasma is most commonly converted from the gas state at extremely high temperatures.

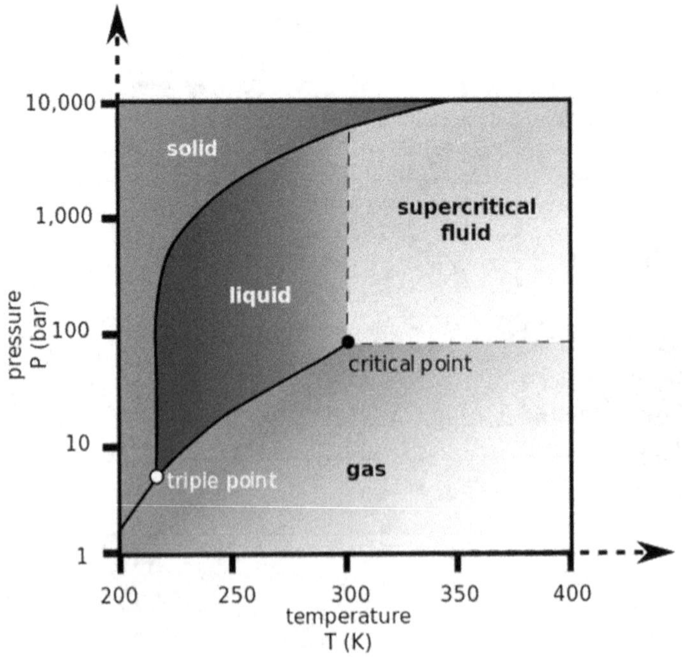

The amount of energy needed to change matter from one state to another is labeled by the terms for phase changes. For example, the temperature needed to supply enough energy for matter to change from a liquid to a gas is called the *heat of vaporization*. When heat is added to matter in order to cause a change in state, there will be an increase in temperature until the matter is about to change its state. During its transition, all of the added heat is used by the matter to change its state, so there is no increase in temperature. Once the transition is complete, then the added heat will again yield an increase in temperature.

Each state of matter is considered to be a phase, and changes between phases are represented by phase diagrams. These diagrams show the effects of changes in pressure and temperature on matter. The states of matter fall into areas on these charts called *heating curves*.

Chemical Nomenclature, Composition, and Bonding

Simple Compounds and Their Chemical Formulas

Chemical formulas represent the proportion of the number of atoms in a chemical compound. Chemical symbols are used for the elements present and numerical values. Parentheses are also sometimes used to show the number of combinations of the elements in relation to their ionic charges. An element's ionic charge can be determined by its location on the periodic table. This information is then used to correctly combine its atoms in a compound.

For example, the chemical formula for sodium chloride (table salt) is the combination of sodium (Na, ionic charge of +1) and chlorine (Cl, ionic charge of -1). From its placement on the periodic table, the electron valence of an outer shell can be determined: sodium has an ionic charge of +1, while chlorine has an ionic charge of -1. Since these two elements have an equal and opposite amount of charge, they combine in a neutral one-to-one ratio: NaCl. The naming of compounds depends mainly on the second element in a chemical compound. If it is a non-metal (such as chlorine), it is written with an "ide" at the end. The compound NaCl is called "sodium chloride."

If the elements forming a compound do not have equal and opposite ionic charges, there will be an unequal number of each element in the compound to balance the ionic charge. This situation happens with many elements, for example, in the combination of nickel and oxygen into nickel oxide (Ni_2O_3). Nickel has a +3 ionic charge and oxygen has a -2 ionic charge, so when forming a compound, there must be two nickel atoms for every three oxygen atoms (a common factor of 6) to balance the charge of the compound. This compound is called "nickel oxide."

A chemical formula can also be written from a compound's name. For instance, the compound carbon dioxide is formed by the combination of carbon and oxygen. The word "dioxide" means there are two oxygen atoms for every carbon atom, so it is written as CO_2.

To better represent the composition of compounds, structural formulas are used. The combination of atoms is more precisely depicted by lining up the electron configuration of the outer electron shell through a Lewis dot diagram.

The Lewis dot diagram, named for Gilbert N. Lewis, shows the arrangement of the electrons in the outer shell and how these electrons can pair/bond with the outer shell electrons of other atoms when forming compounds. The diagram is created by writing the symbol of an element and then drawing dots to represent the outer shell of valence electrons around what would be an invisible square surrounding the symbol. The placement of the first two dots can vary based on the school of teaching. For the given example, the first dot is placed on the top and then the next dot is placed beside it, since it represents the pair of electrons in the 1s valence shell. The next dots (electrons) are placed one at a time on each side—right, bottom, left, right bottom left, etc.—of the element symbol until all of the valence shell electrons are represented, or the structure has eight dots (electrons), which means it is full. This method gives a more specific picture of compounds, how they are structured, and what electrons are available for bonding, sharing, and forming new compounds. For example, the compound sodium chloride is written separately with sodium having one valence electron and chlorine having seven valence electrons. Then, combined with a total of eight electrons, it is written with two dots being shared between the two elements.

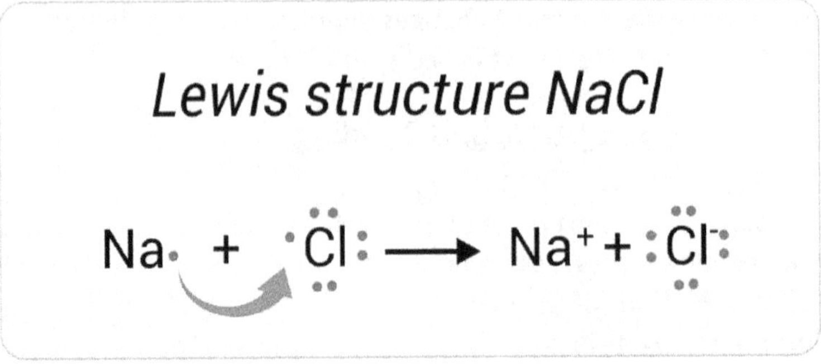

Types of Chemical Bonding

A chemical bond is a strong attractive force that can exist between atoms. The bonding of atoms is separated into two main categories. The first category, *ionic bonding,* primarily describes the bonding that occurs between oppositely charged ions in a regular crystal arrangement. It primarily exists between salts, which are known to be ionic. Ionic bonds are held together by the electrostatic attraction between oppositely charged ions. This type of bonding involves the transfer of electrons from the valence shell of one atom to the valence shell of another atom. If an atom loses an electron from its valence shell, it becomes a positive ion, or *cation*. If an atom gains an electron, it becomes a negative ion, or an *anion*. The Lewis electron-dot symbol is used to more simply express the electron configuration of atoms, especially when forming bonds.

The second type of bonding is covalent bonding. This bonding involves the sharing of a pair of electrons between atoms. There are no ions involved in covalent bonding, but the force holding the atoms together comes from the balance between the attractive and repulsive forces involving the shared electron and the nuclei. Atoms frequently engage is this type of bonding when it enables them to fill their outer valence shell.

Mole Concept and Its Applications

The calculation of mole ratios of reactants and products involved in a chemical reaction is called "stoichiometry." To find these ratios, one must first find the proportion of the number of molecules in one mole of a substance. This relates the molar mass of a compound to its mass and this relationship is a constant known as *Avogadro's number* (6.23×10^{23}). Since it is a ratio, there are no dimensions (or units) for Avogadro's number.

Molar Mass and Percent Composition

The molar mass of a substance is the measure of the mass of one mole of the substance. For pure elements, the molar mass is also known as the atomic mass unit (amu) of the substance. For compounds, it can be calculated by adding the molar masses of each substance in the compound. For example, the molar mass of carbon is 12.01 g/mol, while the molar mass of water (H_2O) requires finding the sum of the molar masses of the constituents ((1.01 x 2 = 2.02 g/mol for hydrogen) + (16.0 g/mol for oxygen) = 18.02 g/mol).

The percentage of a compound in a composition can be determined by taking the individual molar masses of each component divided by the total molar mass of the compound, multiplied by 100. Determining the percent composition of carbon dioxide (CO_2) first requires the calculation of the molar mass of CO_2.

molar mass of carbon = 12.01 x 1 atom = 12.01 g/mol

molar mass of oxygen = 16.0 × 2 atoms = 32.0 g/mol

molar mass of CO_2 = 12.01 g/mol + 32.0 g/mol = 44.01 g/mol

Next, each individual mass is divided by the total mass and multiplied by 100 to get the percent composition of each component.

12.01/44.01 = (0.2729 × 100) = 27.29% carbon

32.0/44.01 = (0.7271 × 100) = 72.71% oxygen

(A quick check in the addition of the percentages should always yield 100%.)

Chemical Reactions

Basic Concepts of Chemical Reactions

Chemical reactions rearrange the initial atoms of the reactants into different substances. These types of reactions can be expressed through the use of balanced chemical equations. A *chemical equation* is the symbolic representation of a chemical reaction through the use of chemical terms. The reactants at the beginning (or on the left side) of the equation must equal the products at the end (or on the right side) of the equation.

For example, table salt (NaCl) forms through the chemical reaction between sodium (Na) and chlorine (Cl) and is written as: $Na + Cl_2 \rightarrow NaCl$.

However, this equation is not balanced because there are two sodium atoms for every pair of chlorine atoms involved in this reaction. So, the left side is written as: $2Na + Cl_2 \rightarrow NaCl$.

Next, the right side needs to balance the same number of sodium and chlorine atoms. So, the right side is written as: $2Na + Cl_2 \rightarrow 2NaCl$. Now, this is a balanced chemical equation.

Chemical reactions typically fall into two types of categories: *endothermic* and *exothermic*.

An endothermic reaction absorbs heat, whereas an exothermic reaction releases heat. For example, in an endothermic reaction, heat is drawn from the container holding the chemicals, which cools the container. Conversely, an exothermic reaction emits heat from the reaction and warms the container holding the chemicals.

Factors that can affect the rate of a reaction include temperature, pressure, the physical state of the reactants (e.g., surface area), concentration, and catalysts/enzymes.

The formula *PV* = *nRT* shows that an increase in any of the variables (pressure, volume, or temperature) affects the overall reaction. The physical state of two reactants can also determine how much interaction they have with each other. If two reactants are both in a fluid state, they may have the capability of interacting more than if solid. The addition of a catalyst or an enzyme can increase the rate of a chemical reaction, without the catalyst or enzyme undergoing a change itself.

Le Chatelier's principle describes factors that affect a reaction's equilibrium. Essentially, when introducing a "shock" to a system (or chemical reaction), a positive feedback/shift in equilibrium is often the response. In accordance with the second law of thermodynamics, this imbalance will eventually even itself out, but not without counteracting the effects of the reaction.

There are many different types of chemical reactions. A *synthesis reaction* is the combination of two or more elements into a compound. For example, the synthesis reaction of hydrogen and oxygen forms water.

$$2\ H_2(g) + O_2(g) \rightarrow 2\ H_2O(g)$$

A *decomposition reaction* is the breaking down of a compound into its more basic components. For example, the decomposition, or electrolysis, of water results in it breaking down into oxygen and hydrogen gas.

$$2\ H_2O \rightarrow 2\ H_2 + O_2$$

A *combustion reaction* is similar to a decomposition reaction, but it requires oxygen and heat for the reaction to occur. For example, the burning of a candle requires oxygen to ignite and the reaction forms carbon dioxide during the process.

$$CH_4(g) + 2\ O_2(g) \rightarrow CO_2(g) + 2\ H_2O(g)$$

There are also single and double replacement reactions where compounds swap components with each other to form new compounds. In the *single replacement reaction*, a single element will swap into a compound, thus releasing one of the compound's elements to become the new single element. For example, the reaction between iron and copper sulfate will create copper and iron sulfate.

$$1\ \mathrm{Fe}(s) + 1\ \mathrm{CuSO_4}(aq) \rightarrow 1\ \mathrm{FeSO_4}(aq) + 1\ \mathrm{Cu}(s)$$

In a *double replacement reaction*, two compounds swap components to form two new compounds. For example, the reaction between sodium sulfide and hydrochloric acid forms sodium chloride and hydrogen sulfide.

$$\mathrm{Na_2S} + \mathrm{HCl} \rightarrow \mathrm{NaCl} + \mathrm{H_2S}$$

After balancing the reaction, we get:

$$\mathrm{Na_2S} + 2\ \mathrm{HCl} \rightarrow 2\ \mathrm{NaCl} + \mathrm{H_2S}$$

An organic reaction is a chemical reaction involving the components of carbon and hydrogen.

Finally, there are oxidation/reduction (redox or half) reactions. These reactions involve the loss of electrons from one species (oxidation), and the gain of electrons to the other species (reduction).

For example, the oxidation of magnesium is as follows:

$$2\ \mathrm{Mg}(s) + \mathrm{O_2}(g) \rightarrow 2\ \mathrm{MgO}(s)$$

Acid-Base Chemistry

Simple Acid-Base Chemistry

If something has a sour taste, it is acidic, and if something has a bitter taste, it is basic. Unfortunately, it can be extremely dangerous to ingest chemicals in an attempt to classify them as an acid or a base. Therefore, acids and bases are generally identified by the reactions they have when combined with water. An acid will increase the concentration of the hydrogen ion (H^+), while a base will increase the concentration of the hydroxide ion (OH^-).

To better categorize the varying strengths of acids and bases, the pH scale is used. The pH scale provides a logarithmic (base 10) grading to acids and bases based on their strength. The pH scale contains values from 0 through 14, with 7 being neutral. If a solution registers below 7 on the pH scale, it is considered an acid. If it registers higher than 7, it is considered a base. To perform a quick test on a solution, litmus paper can be used. A base will turn red litmus paper blue, whereas an acid will turn blue litmus paper red. To gauge the strength of an acid or base, a test of phenolphthalein can be used. An acid will turn red phenolphthalein colorless, and a base will turn colorless phenolphthalein pink. As demonstrated with these types of tests, acids and bases neutralize each other. When acids and bases react with one another, they produce salts (also called ionic substances).

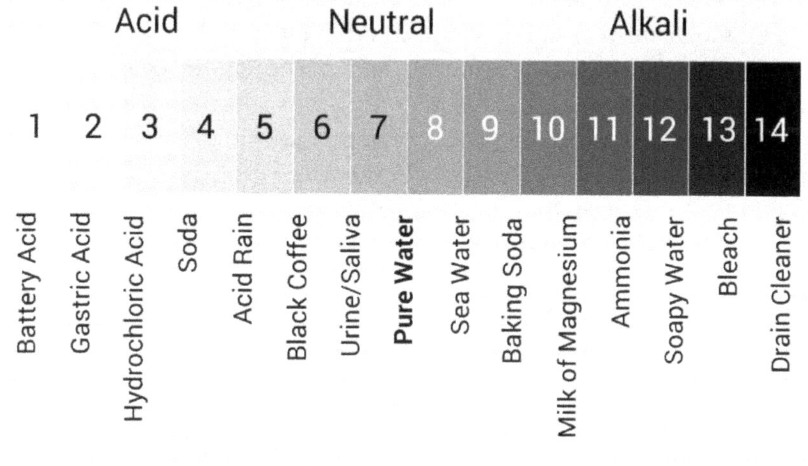
Acid
Neutral
Alkali
1 2 3 4 5 6 7 8 9 10 11 12 13 14
Battery Acid
Gastric Acid
Hydrochloric Acid
Soda
Acid Rain
Black Coffee
Urine/Saliva
Pure Water
Sea Water
Baking Soda
Milk of Magnesium
Ammonia
Soapy Water
Bleach
Drain Cleaner

Solutions and Solubility

Different Types of Solutions

A *solution* is a homogenous mixture of more than one substance. A *solute* is another substance that can be dissolved into a substance called a *solvent*. If only a small amount of solute is dissolved in a solvent, the solution formed is said to be *diluted*. If a large amount of solute is dissolved into the solvent, then the solution is said to be *concentrated*. For example, water from a typical, unfiltered household tap is diluted because it contains other minerals in very small amounts.

Solution Concentration

A
Dilute solution

B
Concentrated solution

If more solute is being added to a solvent, but not dissolving, the solution is called *saturated*. For example, when hummingbirds eat sugar-water from feeders, they prefer it as sweet as possible. When trying to dissolve enough sugar (solute) into the water (solvent), there will be a point where the sugar crystals will no longer dissolve into the solution and will remain as whole pieces floating in the water. At this point, the solution is considered saturated and cannot accept more sugar. This level, at which a solvent cannot accept and dissolve any more solute, is called its *saturation point*. In some cases, it is possible to force more solute to be dissolved into a solvent, but this will result in crystallization. The state of a solution on the verge of crystallization, or in the process of crystallization, is called a *supersaturated* solution. This can also occur in a solution that seems stable, but if it is disturbed, the change can begin the crystallization process.

Although the terms *dilute, concentrated, saturated*, and *supersaturated* give qualitative descriptions of solutions, a more precise quantitative description needs to be established for the use of chemicals. This holds true especially for mixing strong acids or bases. The method for calculating the concentration of a solution is done through finding its molarity. In some instances, such as environmental reporting, molarity is measured in parts per million (ppm). Parts per million, is the number of milligrams of a substance dissolved in one liter of water. To find the *molarity*, or the amount of solute per unit volume of solution, for a solution, the following formula is used:

$$c = \frac{n}{V}$$

In this formula, c is the molarity (or unit moles of solute per volume of solution), n is the amount of solute measured in moles, and V is the volume of the solution, measured in liters.

Example:

What is the molarity of a solution made by dissolving 2.0 grams of NaCl into enough water to make 100 mL of solution?

To solve this, the number of moles of NaCl needs to be calculated:

First, to find the mass of NaCl, the mass of each of the molecule's atoms is added together as follows:

23.0g (Na) + 35.5g (Cl) = 58.5g NaCl

Next, the given mass of the substance is multiplied by one mole per total mass of the substance:

2.0g NaCl × (1 mol NaCl/58.5g NaCl) = 0.034 mol NaCl

Finally, the moles are divided by the number of liters of the solution to find the molarity:

(0.034 mol NaCl)/(0.100L) = 0.34 M NaCl

To prepare a solution of a different concentration, the *mass solute* must be calculated from the molarity of the solution. This is done via the following process:

Example:

How would you prepare 600.0 mL of 1.20 M solution of sodium chloride?

To solve this, the given information needs to be set up:

1.20 M NaCl = 1.20 mol NaCl/1.00 L of solution

0.600 L solution × (1.20 mol NaCl/1.00 L of solution) = 0.72 moles NaCl

0.72 moles NaCl × (58.5g NaCl/1 mol NaCl) = 42.12 g NaCl

This means that one must dissolve 42.12 g NaCl in enough water to make 600.0 L of solution.

Factors Affecting the Solubility of Substances and the Dissolving Process

Certain factors can affect the rate in dissolving processes. These include temperature, pressure, particle size, and agitation (stirring). As mentioned, the *ideal gas law* states that $PV = nRT$, where P equals pressure, V equals volume, and T equals temperature. If the pressure, volume, or temperature are affected in a system, it will affect the entire system. Specifically, if there is an increase in temperature, there will be an increase in the dissolving rate. An increase in the pressure can also increase the dissolving rate. Particle size and agitation can also influence the dissolving rate, since all of these factors contribute to the breaking of intermolecular forces that hold solute particles together. Once these forces are broken, the solute particles can link to particles in the solvent, thus dissolving the solute.

A *solubility curve* shows the relationship between the mass of solute that a solvent holds at a given temperature. If a reading is on the solubility curve, the solvent is *full* (*saturated*) and cannot hold anymore solute. If a reading is above the curve, the solvent is *unstable* (*supersaturated*) from holding more solute than it should. If a reading is below the curve, the solvent is *unsaturated* and could hold more solute.

If a solvent has different electronegativities, or partial charges, it is considered to be *polar*. Water is an example of a polar solvent. If a solvent has similar electronegativities, or lacking partial charges, it is considered to be *non-polar*. Benzene is an example of a non-polar solvent. Polarity status is important when attempting to dissolve solutes. The phrase "like dissolves like" is the key to remembering what will happen when attempting to dissolve a solute in a solvent. A polar solute will dissolve in a like, or polar solvent. Similarly, a non-polar solute will dissolve in a non-polar solvent. When a reaction produces a solid, the solid is called a *precipitate.* A precipitation reaction can be used for removing a salt (an ionic compound that results from a neutralization reaction) from a solvent, such as water. For water, this process is called ionization. Therefore, the products of a neutralization reaction (when an acid and base react) are a salt and water. Therefore, the products of a neutralization reaction (when an acid and base react) are a salt and water.

When a solute is added to a solvent to lower the freezing point of the solvent, it is called *freezing point depression*. This is a useful process, especially when applied in colder temperatures. For example, the addition of salt to ice in winter allows the ice to melt at a much lower temperature, thus creating safer road conditions for driving. Unfortunately, the freezing point depression from salt can only lower the melting point of ice so far and is ineffectual when temperatures are too low. This same process, with a mix of ethylene glycol and water, is also used to keep the radiator fluid (antifreeze) in an automobile from freezing during the winter.

Cell Biology

Function of Cells

Structure and Function of Cell Membranes

All cells are surrounded by a cell membrane that is formed from two layers of phospholipids. *Phospholipids* are two fatty acid chains connected to a glycerol molecule with a phosphate group. The membrane is amphiphilic because the fatty acid chains are hydrophobic and the phosphate group is hydrophilic. This creates a unique environment that protects the cell's inner contents while still allowing material to pass through the membrane. Because the outside of a cell, known as the *extracellular space*, and the inside of a cell, the *intercellular space*, are aqueous, the lipid bilayer forms with the two phospholipid heads facing the outside and the inside of the cell. This allows the phospholipids to interact with water; the fatty acid tails face the middle, so they can interact with each other and avoid water.

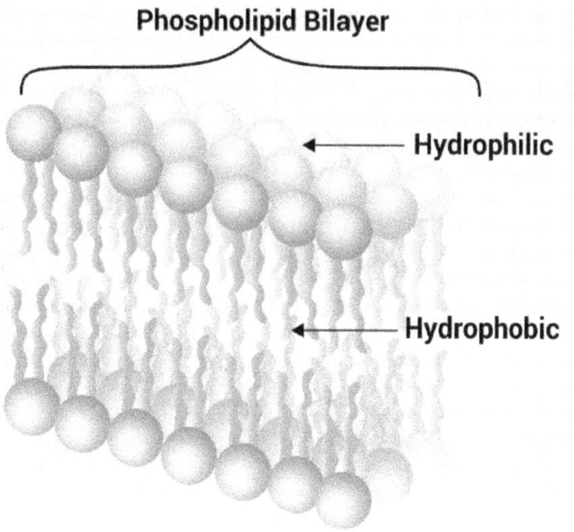

Molecules can pass through the cell membrane by either active or passive transport. *Active transport* requires chemical energy to move molecules in or out of the cell from areas of low concentration to areas of high concentration, or in instances where the molecules would not be able to pass through by themselves, such as with large non-lipid soluble molecules. Ions, amino acids, and complex sugars use active transport mechanisms. Molecules that are soluble in lipids, water, and oxygen use *passive transport* to move in and out of the cell, which means that cellular energy is not required for their movement. Examples of passive transport include diffusion, facilitated diffusion, and osmosis. *Diffusion* is the net movement of particles from an area of high concentration to lower concentration. *Facilitated diffusion* is the movement of molecules through cell membranes with the use of special transport proteins. Finally, *osmosis* is the movement of water molecules across partially permeable membranes.

Structure and Function of Animal and Plant Cell Organelles

Animal and plant cells contain many of the same or similar *organelles*, which are membrane enclosed structures that each have a specific function; however, there are a few organelles that are unique to either one or the other general cell type. The following cell organelles are found in both animal and plant cells, unless otherwise noted in their description:

- *Nucleus*: The nucleus consists of three parts: the nuclear envelope, the nucleolus, and chromatin. The *nuclear envelope* is the double membrane that surrounds the nucleus and separates its contents from the rest of the cell. The *nucleolus* produces ribosomes. *Chromatin* consists of DNA and protein, which form chromosomes that contain genetic information. Most cells have only one nucleus; however, some cells, such as skeletal muscle cells, have multiple nuclei.

- *Endoplasmic reticulum (ER)*: The ER is a network of membranous sacs and tubes that is responsible for membrane synthesis. It is also responsible for packaging and transporting proteins into vesicles that can move out of the cell. It folds and transports other proteins to the Golgi apparatus. It contains both smooth and rough regions; the rough regions have ribosomes attached, which are the sites of protein synthesis.

- *Flagellum*: Flagella are found only in animal cells. They are made up of a cluster of microtubules projected out of the plasma membrane, and they aid in cell mobility.

- *Centrosome*: The centrosome is the area of the cell where *microtubules*, which are filaments that are responsible for movement in the cell, begin to be formed. Each centrosome contains two centrioles. Each cell contains one centrosome.

- *Cytoskeleton*: The cytoskeleton in animal cells is made up of microfilaments, intermediate filaments, and microtubules. In plant cells, the cytoskeleton is made up of only microfilaments and microtubules. These structures reinforce the cell's shape and aid in cell movement.

- *Microvilli*: Microvilli are found only in animal cells. They are protrusions in the cell membrane that increase the cell's surface area. They have a variety of functions, including absorption, secretion, and cellular adhesion. They are found on the apical surface of epithelial cells, such as in the small intestine. They are also located on the plasma surface of a female's eggs to help anchor sperm that are attempting fertilization.

- *Peroxisome*: A peroxisome contains enzymes that are involved in many of the cell's metabolic functions, one of the most important being the breakdown of very long chain fatty acids. Peroxisomes produces hydrogen peroxide as a byproduct of these processes and then converts the hydrogen peroxide to water. There are many peroxisomes in each cell.

- *Mitochondrion*: The mitochondrion is often called the powerhouse of the cell and is one of the most important structures for maintaining regular cell function. It is where aerobic cellular respiration occurs and where most of the cell's adenosine triphosphate (ATP) is generated. The number of mitochondria in a cell varies greatly from organism to organism, and from cell to cell. In human cells, the number of mitochondria can vary from zero in a red blood cell, to 2000 in a liver cell.

- *Lysosome*: Lysosomes are responsible for digestion and can hydrolyze macromolecules. There are many lysosomes in each cell.

- *Golgi apparatus*: The Golgi apparatus is responsible for the composition, modification, organization, and secretion of cell products. Because of its large size, it was actually one of the first organelles to be studied in detail. There are many Golgi apparatuses in each cell.
- *Ribosomes*: Ribosomes are found either free in the cytosol, bound to the rough ER, or bound to the nuclear envelope. They manufacture proteins within the cell.
- *Plasmodesmata*: The plasmodesmata are found only in plant cells. They are cytoplasmic channels, or tunnels, that go through the cell wall and connect the cytoplasm of adjacent cells.
- *Chloroplast*: Chloroplasts are found only in plant cells. They are responsible for *photosynthesis*, which is the process of converting sunlight to chemical energy that can be stored and used later to drive cellular activities.

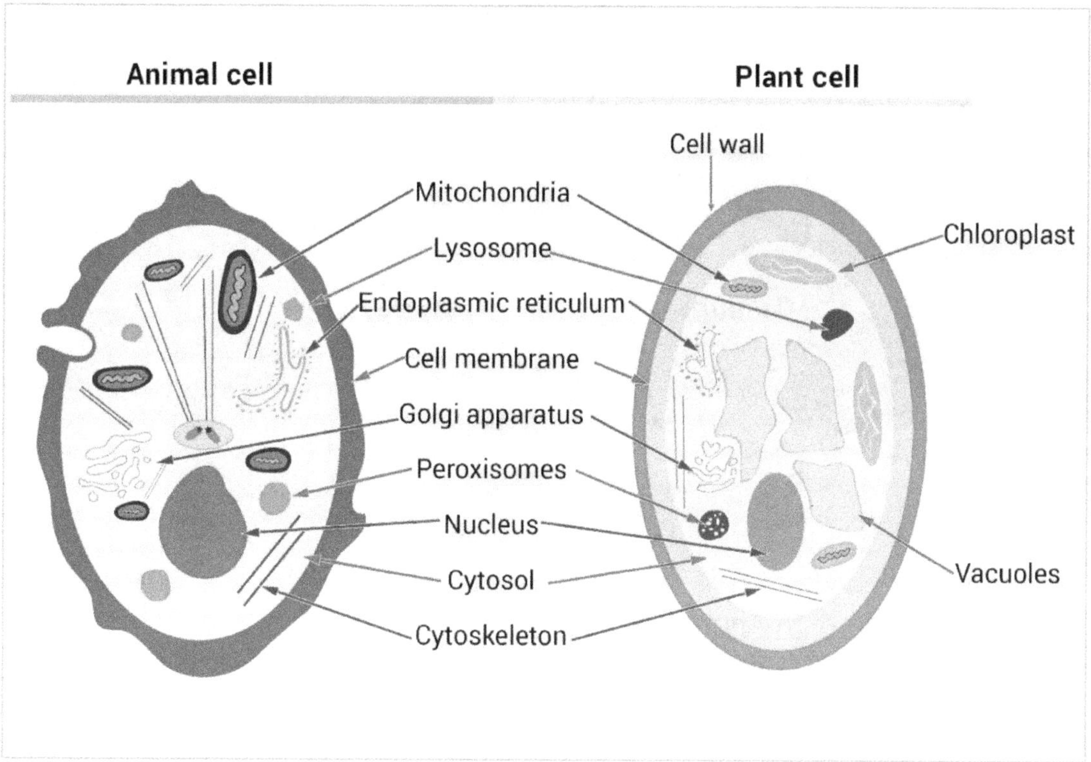

- *Central vacuole*: A central vacuole is found only in plant cells. It is responsible for storing material and waste. This is the only vacuole found in a plant cell.
- *Plasma membrane*: The plasma membrane is a phospholipid bilayer that encloses the cell.
- *Cell wall*: Cell walls are only present in plant cells. The cell wall is made up of strong fibrous substances including cellulose and other polysaccharides, and protein. It is a layer outside of the plasma membrane, which protects the cell from mechanical damage and helps maintain the cell's shape.

Levels of Organization

There are about two hundred different types of cells in the human body. Cells group together to form *biological tissues*, and tissues combine to form *organs*, such as the heart and kidneys. Organs that work together to perform vital functions of the human body form *organ systems*. There are eleven organ systems in the human body: skeletal, muscular, urinary, nervous, digestive, endocrine, reproductive, respiratory, cardiovascular, integumentary, and lymphatic. Although each system has its own unique function, they all rely on each other, either directly or indirectly, to operate properly.

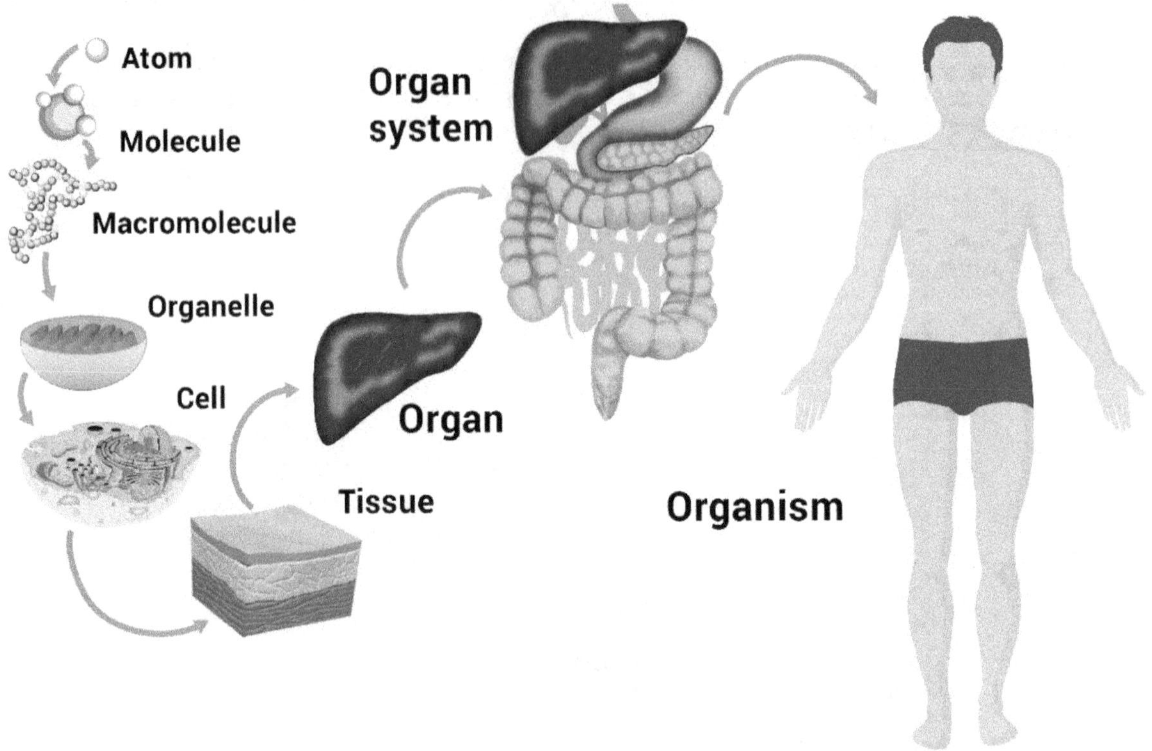

Major Features of Common Animal Cell Types

The most common animal cell types are blood, muscle, nerve, epithelial, and gamete cells. The three main blood cells are *red blood cells (RBCs), white blood cells (WBCs),* and *platelets*. RBCs transport oxygen and carbon dioxide through the body. They do not have a nucleus and they live for about 120 days in the blood. WBCs defend the body against diseases. They do have a nucleus and live for only three to four days in the human body. Platelets help with the formation of blood clots following an injury. They do not have a nucleus and live for about eight days after formation.

Muscle cells are long, tubular cells that form muscles, which are responsible for movement in the body. On average, they live for about fifteen years, but this number is highly dependent on the individual body. There are three main types of muscle tissue: skeletal, cardiac, and smooth. *Skeletal muscle cells* have multiple nuclei and are the only voluntary muscle cell, which means that the brain consciously controls the movement of skeletal muscle. *Cardiac muscle cells* are only found in the heart; they have a single nucleus and are involuntary. *Smooth muscle cells* make up the walls of the blood vessels and organs. They have a single nucleus and are involuntary.

Nerve cells conduct electrical impulses that help send information and instructions from the brain to the rest of the body. They contain a single nucleus and have a specialized membrane that allows for this electrical signaling between cells.

Epithelial cells cover exposed surfaces, and line internal cavities and passageways.

Gametes are specialized cells that are responsible for reproduction. In the human body, the gametes are the egg and the sperm.

Prokaryotes and Eukaryotes

There are two distinct types of cells that make up most living organisms: *prokaryotic* and *eukaryotic*. Both types of cells are enclosed by a cell membrane, which is selectively permeable. Selective permeability means essentially that it is a gatekeeper, allowing certain molecules and ions in and out, and keeping unwanted ones at bay, at least until they are ready for use. They both contain *ribosomes* and DNA. One major difference between these types of cells is that in eukaryotic cells, the cell's DNA is enclosed in a membrane-bound nucleus, whereas in prokaryotic cells, the cell's DNA is in a region—called the *nucleoid*—that is not enclosed by a membrane. Another major difference is that eukaryotic cells contain organelles, while prokaryotic cells do not have organelles.

Prokaryotic cells include *bacteria* and archaea. They do not have a nucleus or any membrane-bound organelles, are unicellular organisms, and are generally very small in size. Eukaryotic cells include animal, plant, fungus, and protist cells. *Fungi* are unicellular microorganisms such as yeasts, molds, and mushrooms. Their distinguishing characteristic is the chitin that is in their cell walls. *Protists* are organisms that are not classified as animals, plants, or fungi; they are unicellular; and they do not form tissues.

Key Aspects of Cell Reproduction and Division

Cell Cycle

The *cell cycle* is the process by which a cell divides and duplicates itself. There are two processes by which a cell can divide itself: mitosis and meiosis. In *mitosis*, the daughter cells that are produced from parental cell division are identical to each other and the parent.

Meiosis is a unique process that involves two stages of cell division and produces *haploid cells*, which are cells containing only one set of chromosomes, from *diploid parent cells*, which are cells containing two sets of chromosomes.

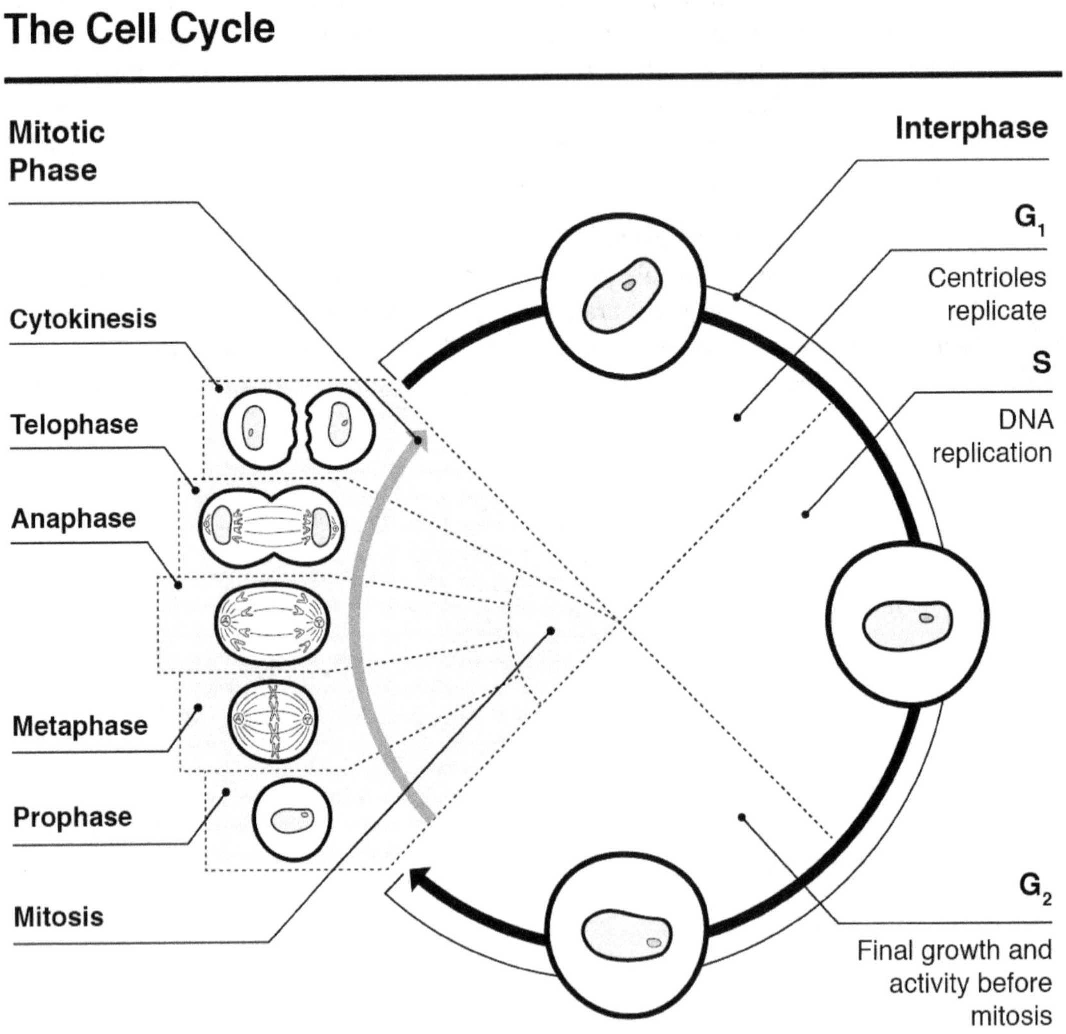

Mitosis

Mitosis can be broken down into five stages: prophase, prometaphase, metaphase, anaphase, and telophase.

- *Prophase*: During this phase, the mitotic spindles begin to form from centrosomes and microtubules. As the microtubules lengthen, the centrosomes move farther away from each other. The nucleolus disappears and the chromatin fibers begin to coil up and form chromosomes. Two sister chromatids, which are two copies of one chromosome, are joined together.

- *Prometaphase*: The nuclear envelope begins to break down and the microtubules enter the nuclear area. Each pair of chromatin fibers develops a *kinetochore*, which is a specialized protein structure in the middle of the adjoined fibers. The chromosomes are further condensed.

- *Metaphase*: In this stage, the microtubules are stretched across the cell and the centrosomes are at opposite ends of the cell. The chromosomes align at the *metaphase plate*, which is a plane that is exactly between the two centrosomes. The kinetochore of each chromosome is attached to the kinetochore of the microtubules that are stretching from each centrosome to the metaphase plate.

- *Anaphase*: The sister chromatids break apart, forming full-fledged chromosomes. The two daughter chromosomes move to opposite ends of the cell. The microtubules shorten toward opposite ends of the cell as well, and the cell elongates.

- *Telophase*: Two nuclei form at each end of the cell and nuclear envelopes begin to form around each nucleus. The nucleoli reappear and the chromosomes become less condensed. The microtubules are broken down by the cell and mitosis is complete.

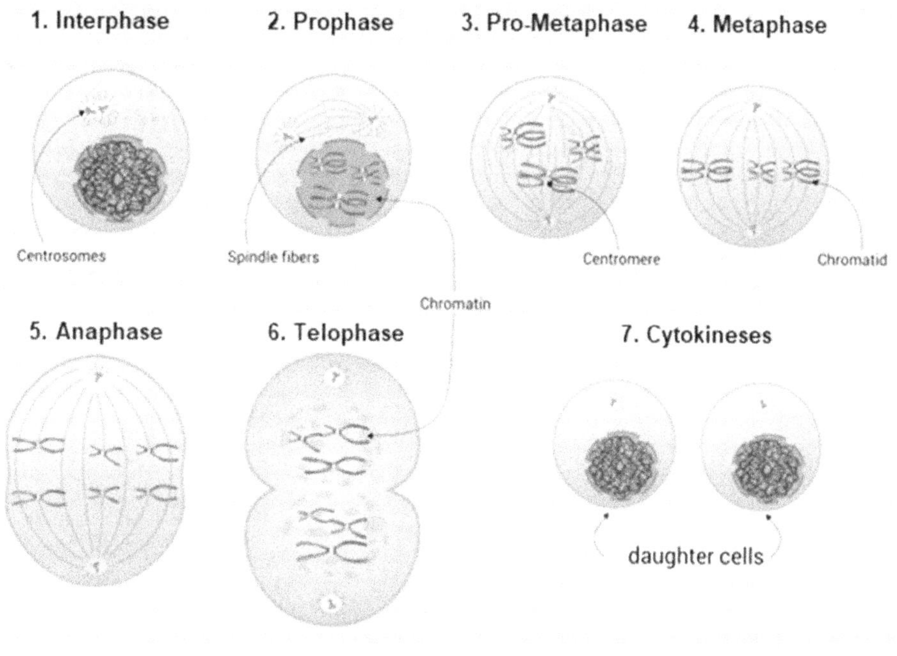

Meiosis

Meiosis is a type of cell division in which the daughter cells have half as many sets of chromosomes as the parent cell. In addition, one parent cell produces four daughter cells. Meiosis has the same phases as mitosis, except that they occur twice—once in meiosis I and once in meiosis II. The diploid parent has two sets of chromosomes, set A and set B. During meiosis I, each chromosome set duplicates, producing a second set of A chromosomes and a second set of B chromosomes, and the cell splits into two. Each cell contains two sets of chromosomes. Next, during meiosis II, the two intermediate daughter cells divide again, producing four total haploid cells that each contain one set of chromosomes. Two of the haploid cells each contain one chromosome of set A and the other two cells each contain one chromosome of set B.

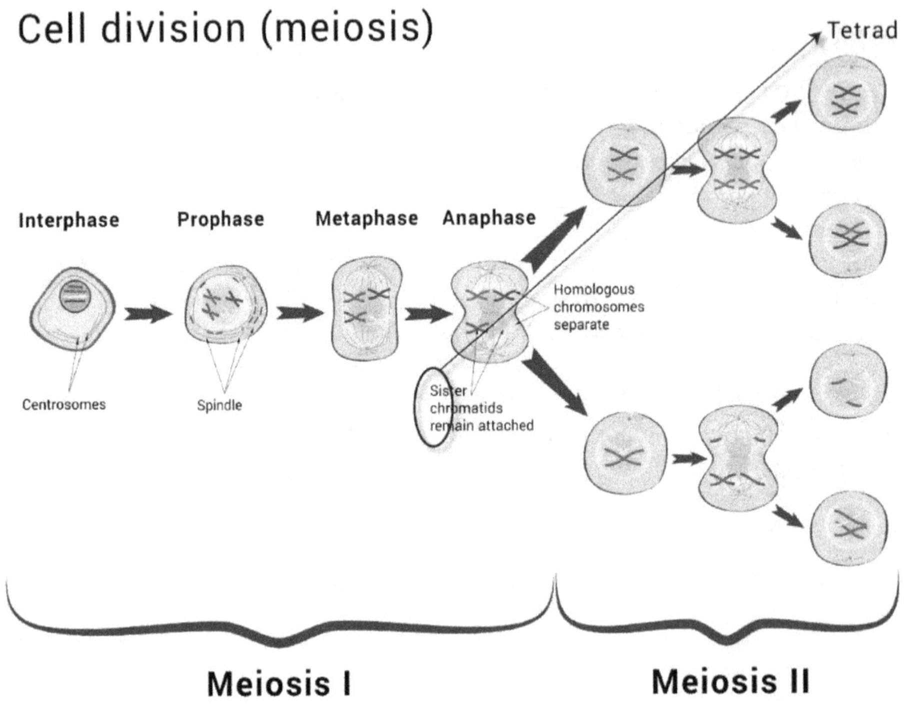

Cytokinesis

Cytokinesis is the division of cytoplasm that occurs immediately following the division of genetic material during cellular reproduction. The process of mitosis or meiosis, followed by cytokinesis, finishes up the cell cycle.

Histology

Histology

Histology is the examination of specialized cells and cell groups that perform a specific function by working together. Although there are trillions of cells in the human body, there are only 200 different types of cells. Groups of cells form biological tissues, and tissues combine to form organs, such as the heart and kidney. Organs are structures that have many different functions that are vital to living creatures. There are four primary types of tissue: epithelial, connective, muscle, and neural. Each tissue type has specific characteristics that enable organs and organ systems to function properly.

Epithelial Tissue

Epithelial tissue includes epithelia and glands. Epithelia are the layers of cells that cover exposed surfaces and line internal cavities and passageways. Glands are structures that are involved in secretion of fluids. Epithelia do not contain blood vessels and can often regenerate quickly to replace dead and damaged cells.

Connective Tissue

Connective tissue fills internal spaces and is not exposed to the outside of the body. It provides structural support for the body and stores energy. Fluid can be transported by connective tissue between different regions of the body. This type of tissue is also a protective barrier for delicate organs and for the body against microorganisms. The cells in epithelial tissue are tightly packed together. However, those in connective tissue are comparatively sparse and scattered in the extracellular matrix which is composed of glycoproteins and fibrous proteins connected to a basement membrane.

The body contains many types of connective tissue. Loose connective tissue is the most common form; it adheres connective tissue to underlying tissue and holds organs in place. There are three main types of loose connective tissue fibers—collagenous, elastic, and reticular. Dense, or fibrous, connective tissue is found in tendons, ligaments, and in the dermis of the skin. There are also specialized connective tissues such as adipose (fat), bone, cartilage, blood, and lymph.

Muscle Tissue

Muscle tissue has characteristics that allow motion. Skeletal muscles are long fibers of actin and myosin that slide past each other to cause a contraction, which shortens the muscle. They are filled with mitochondria, because they expend a lot of energy. Smooth muscle tissue is structured differently, because its movement is in the form of peristalsis. For example, the esophagus moves food in a wave-like contraction, as opposed to a shortening of muscle to bring structures together. Smooth muscle tissues make up internal organs, with the exception of the heart, which is composed of thick, contracting muscle tissue.

Neural Tissue

Neural tissue conducts electrical impulses which help send information and instructions throughout the body. Most of it is concentrated in the brain and spinal cord. The changes in frequency and patterns of the impulse are important distinctions in the messages being sent. Its structure is composed of message-receiving projections called dendrites and a long, myelinated axon. The myelin sheath surrounds the axon to conduct the action potential, ultimately causing neurotransmitter release at the synapse, or the boundary between the axon of one cell and the dendrite of another.

Integumentary System

The integumentary system includes skin, hair, nails, oil glands, and sweat glands. The largest organ of the integumentary system (and of the body), the skin, acts as a barrier and protects the body from mechanical impact, variations in temperature, microorganisms, chemicals, and UV radiation from the sun. It regulates body temperature, peripheral circulation, and excretes waste through sweat. It also contains a large network of nerve cells that relay changes in the external environment to the brain.

Layers of Skin

Skin consists of three layers: the surface epidermis, the inner dermis, and the subcutaneous hypodermis that is below the dermis and contains a layer of fat and connective tissue, which are both important for insulation.

The whole epidermis is composed of epithelial cells that lack blood vessels. The outer epidermis is composed of dead cells, which surround the living cells underneath. The most inner epidermal tissue is a single layer of cells called the stratum basale, which is composed of rapidly dividing cells that push old cells to the skin's surface. When being pushed out, the cells' organelles disappear, and they start producing a protein called keratin, which eventually forms a tough waterproof layer. This outer layer sloughs off every four to five weeks. The melanocytes in the stratum basale produce the pigment melanin, which absorbs UV rays and protects the skin. Skin also produces vitamin D if exposed to sunlight.

The dermis underneath the epidermis contains supporting collagen fibers peppered with nerves, blood vessels, hair follicles, sweat glands, oil glands, and smooth muscles.

Skin's Involvement in Temperature Homeostasis

The skin has a thermoregulatory role in the human body that is controlled by a negative feedback loop. The control center of temperature regulation is the hypothalamus in the brain. When the hypothalamus is alerted by receptors from the dermis, it secretes hormones that activate effectors to keep internal temperature at a set point of 98.6°F (37°C). If the environment is too cold, the hypothalamus will initiate a pathway that induces muscle shivering to release heat energy as well as constrict blood vessels to limit heat loss. In hot conditions, the hypothalamus will initiate a pathway that vasodilates blood vessels to increase heat loss and stimulate sweating for evaporative cooling. Evaporative cooling occurs when the hottest water particles evaporate and leave behind the coolest ones. This cools down the body.

Sebaceous Glands vs. Sweat Glands

The skin also contains oil glands, or sebaceous glands, and sweat glands that are exocrine because their substances are secreted through ducts. Endocrine glands secrete substances into the blood stream instead. Oil glands are attached to hair follicles. They secrete sebum, an oily substance that moisturizes the skin, protecting it from water loss. Sebum also keeps the skin elastic. Also, sebum's slight acidity provides a chemical defense against bacterial and fungal infections.

Sweat glands not attached to hair follicles are called eccrine glands. They are all over the body, and are the sweat glands responsible for thermoregulation. They also remove bodily waste by secreting water and electrolytes. Sweat glands attached to hair follicles are apocrine glands, and there are far fewer apocrine sweat glands than eccrine sweat glands in the body. Apocrine glands are only active postpuberty. They secrete a thicker, viscous substance that is attractive to bacteria, leading to the unpleasant smell in armpits, feet, and the groin. They are stimulated during stress and arousal.

Skeletal System & Articulations

Axial Skeleton and Appendicular Skeleton

The skeletal system is composed of 206 bones interconnected by tough connective tissue called ligaments. The axial skeleton can be considered the north-south axis of the skeleton. It includes the spinal column, sternum, ribs, and skull. There are 80 bones in the axial skeleton, and 33 of them are vertebrae. The ribs make up 12 of the bones in the axial skeleton.

The remaining 126 bones are in the appendicular skeleton, which contains bones of the appendages like the collarbone (clavicle), shoulders (scapula), arms, hands, hips, legs, and feet. The arm bones consist of the upper humerus with the radius and ulna that attach to the hands. The wrists, hands, and fingers are composed of the carpals, metacarpals, and phalanges, respectively. The femur attaches to the hips. The patella or kneecap connects the femur to the fibula and tibia. The ankles, feet, and toes are composed of the tarsals, metatarsals, and phalanges, respectively.

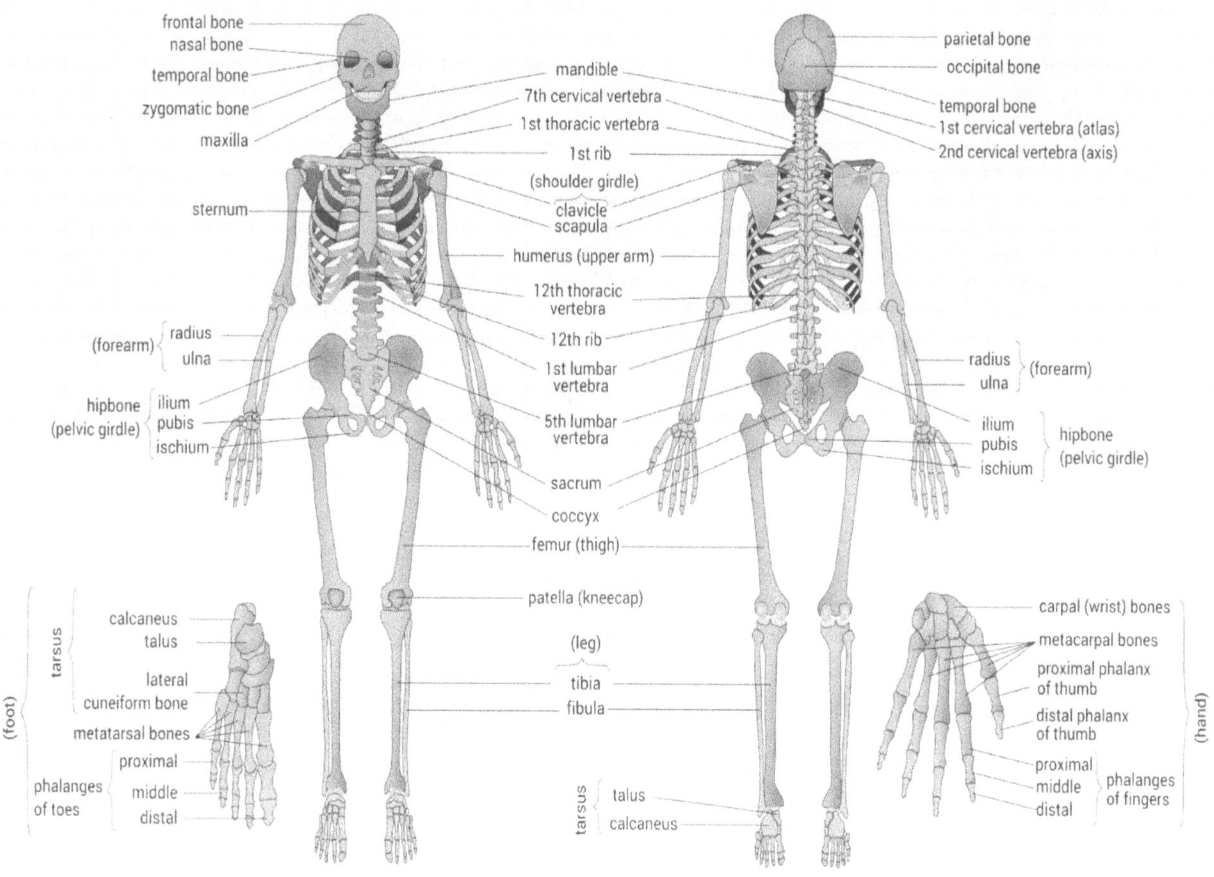

Functions of the Skeletal System

One of the skeletal system's most important functions is to protect vital internal organs. The skull protects the brain, the ribs protect the lungs and heart, the vertebrae protect the spinal column, and the pelvis protects the uterus and other reproductive organs. Red bone marrow produces red and white blood cells as well as platelets in a process known as hemopoiesis. The bones themselves store the essential minerals calcium and phosphorus. The organization of the skeleton allows us to stand upright and acts as a foundation for organs and tissues to attach and maintain their location. This is similar to how the wooden frame of a house has room partitions to designate the type of room and floors that furniture can attach to.

The skeletal system and the muscular system are literally interconnected and allow for voluntary movement. Strong connective tissues called tendons attach bones to muscles. Most muscles work in opposing pairs and act as levers. For example, flexing the biceps brings the arm bones together and flexing the triceps pushes them apart. Synovial joints are movable joints, and they are rich with cartilage, connective tissue, and synovial fluid which acts as a lubricant. The majority of joints are synovial joints, and they include hinge joints, like the one at the elbow, which allows for opening and closing.

The vertebrae are cartilaginous joints, which have spaces between them filled with cushion-like intervertebral discs that act as shock absorbers. The tight fit between the vertebrae and the intervertebral discs helps to protect the spinal cord inside by limiting the movement between two adjacent vertebrae. However, because there are so many vertebrae, the backbone as a whole is somewhat flexible.

Fibrous joints like those in the skull have fibrous tissue between the bones and no cavity between them. These are fixed joints that are immobile.

Compact and Spongy Bone

Osteoclasts, osteoblasts, and osteocytes are the three types of bone cells. Osteoclasts break down old bone, osteoblasts make new bone, and osteocytes are the mature functional bone cells. Bone is constantly regenerating due to the osteoblasts/osteoclasts that line all types of bones and the blood vessels inside them. The cells exist within a matrix of collagen fibers and minerals. The collagen fibers provide resistance to tension and the minerals provide resistance to compression. Because of the collagen and mineral matrix, bones have ample reinforcement to collectively support the entire human body.

Bones can be classified as any of the following:

- Long bones include tube-like rods like the arm and leg bones.
- Short bones are tube-like rods that are smaller than long bones like the fingers and toes.
- Flat bones are thin and flat like the ribs and breastbone.
- Irregular bones, like the vertebrae, are compact and don't fit into the other categories.

The outer tissue of the bone is surrounded by connective tissue known as periosteum. It appears shiny, smooth, and white. It protects the bone, anchors the bone to the connective tissue that surrounds muscles, and links the bone to the circulatory and nervous system. Compact bone is underneath the periosteum and is made of a dense blend of tightly packed osteocytes. It serves as a mineral reservoir of calcium and phosphorus. Compact bones have a Haversian system that is composed of embedded blood vessels, lymph vessels, and nerve bundles that span the interior of the bone from one end to the other. Branching from the central canal to the surface of the bone are the canals of Volkmann which deliver materials to peripheral osteocytes. Concentric circles surround the central Haversian canal, and these lamallae have gaps between them called lacunae where osteocytes are embedded.

In contrast, spongy bone is very porous and more flexible than compact bone. It is at the ends of long bones and the central part of flat bones. It looks like a honeycomb, and the open spaces are connected by trabeculae which are beams of tissue that add support. They add strength without adding mass.

Cartilage is a very flexible connective tissue made of collagen and the flexible elastin. It has no blood vessels and obtains materials via diffusion. It is replaced by bone starting in infancy in a process called ossification.

Joint Structure and Function

Joints can be classified based on structure of how the bones are connected:

- Fibrous joints: Bones joined by fibrous tissue and that lack a joint cavity, e.g., sutures of the skull
- Cartilaginous joints: Bones joined by cartilage and lack a joint cavity, e.g., the pubic symphysis
- Synovial joints: bones separated by a fluid-containing joint cavity with articular cartilage covering the ends of the bone and forming a capsule
- Plane joints: flat surfaces that allow gliding and transitional movements, e.g., intercarpal joints
- Hinge joints: cylindrical projection that nests in a trough-shaped structure, single plane of movement, e.g., the elbow
- Pivot joints: rounded structure that sits into a ring-like shape, allowing uniaxial rotation of the bone around the long axis, e.g., radius head on ulna
- Condyloid joints: oval articular surface that nests in a complementary depression, allowing all angular movements, e.g., the wrist

- Saddle joints: articular surfaces that both have complementary concave and convex areas, allowing more movement than condyloid joints, e.g., the thumb
- Ball-and-socket joints: spherical structure that fits in a cuplike structure, allowing multiaxial movements, e.g., the shoulder

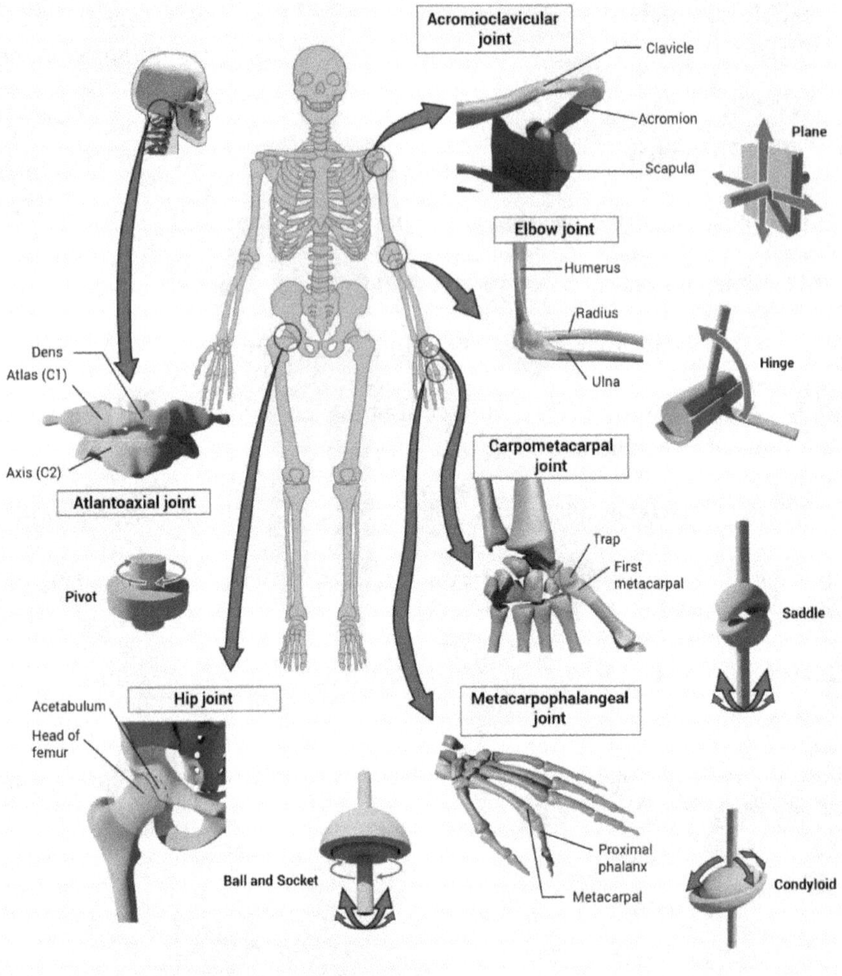

Range of Motion

Range of motion refers to the amount of movement or distance that a joint can travel between the fully flexed and extended positions. Massage therapists and fitness and bodywork professionals use various stretching modalities to try and increase range of motion when it is limited due to tightness in series, elastic, or contractile elements of muscles, tendons, fascia, ligaments, or cartilage surrounding a joint. Inflammation within the joint space can also reduce the available range of motion as can pain and guarding.

Passive range of motion refers to the motion that the joint allows entirely under the therapist's control; the client remains completely relaxed and allows the therapist to manually move the limb segments around the joint to assess the available motion. Passive range of motion only assesses the non-contractile structures surrounding a joint and any hypermobility or limited range of motion can be assessed in structures such as ligaments, joint capsule, and fascia.

With active range of motion, the client uses their muscles to move the joint instead of the therapist providing the mechanical power. In this way, active range of motion assesses both the non-contractile and contractile elements, including the muscles and tendons themselves. If a therapist determines that there is hypomobility at a joint due to the limitations in the contractile components, massage can be useful in bringing blood flow to the area and attempting to break up any adhesions or tightness within the fibers and tissues.

Resisted range of motion is a form of active range of motion testing in which the therapist assesses the strength of the muscles and contractile elements surrounding the joint. The joint is placed in its neutral position. The client is asked to contract the muscles to either flex or extend the joint being evaluated, while the therapist provides resistance in the form of the opposing motion at the joint.

For example, when evaluating biceps flexion strength, the arm should be held at the side in the neutral position. Then the patient is asked to flex the arm, bending the elbow. The therapist should place their hands on the forearm just above the wrist and just below the shoulder on the upper arm, isolating the elbow joint. The therapist should be mindful to never cross two joints at once with their hand position, because this can cause injury. While the client flexes their elbow, the therapist provides an extension force to the elbow, countering flexion and providing resistance. The therapist can gauge the strength of the client's contraction based on the amount of resistance her she can apply to the joint. This type of exercise is contraindicated for any client who has an acute injury to the area being evaluated. The amount of resistance provided by the massage therapist should be appropriate to the health and physical ability of the client; maximal resistance is never recommended.

Muscular System

The muscular system is responsible for involuntary and voluntary movement of the body. There are three types of muscle: skeletal, cardiac, and smooth.

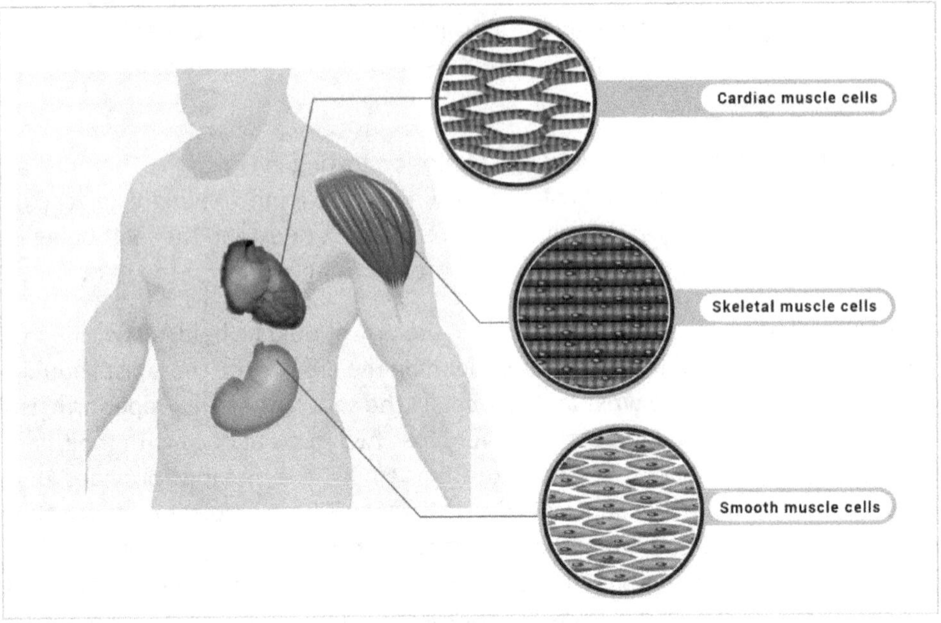

There are approximately 700 muscles in the body that are attached to the bones of the skeletal system and that make up half of the body's weight. Muscles are attached to the bones through tendons. Tendons are made up of dense bands of connective tissue and have collagen fibers that firmly attach to the bone on one side and the muscle on the other. Their fibers are actually woven into the coverings of the bone and muscle so they can withstand the large forces that are put on them when muscles are moving. There are three types of muscle tissue in the body: Skeletal muscle tissue pulls on the bones of the skeleton and causes body movement; cardiac muscle tissue helps pump blood through veins and arteries; and smooth muscle tissue helps move fluids and solids along the digestive tract and contributes to movement in other body systems. All of these muscle tissues have four important properties in common: They are excitable, meaning they respond to stimuli; contractile, meaning they can shorten and pull on connective tissue; extensible, meaning they can be stretched repeatedly, but maintain the ability to contract; and elastic, meaning they rebound to their original length after a contraction.

Muscles begin at an origin and end at an insertion. Generally, the origin is proximal to the insertion and the origin remains stationary while the insertion moves. For example, when bending the elbow and moving the hand up toward the head, the part of the forearm that is closest to the wrist moves and the part closer to the elbow is stationary. Therefore, the muscle in the forearm has an origin at the elbow and an insertion at the wrist.

Body movements occur by muscle contraction. Each contraction causes a specific action. Muscles can be classified into one of three muscle groups based on the action they perform. Primary movers, or agonists, produce a specific movement, such as flexion of the elbow. Synergists are in charge of helping the primary movers complete their specific movements. They can help stabilize the point of origin or provide extra pull near the insertion. Some synergists can aid an agonist in preventing movement at a joint. Antagonists are muscles whose actions are the opposite of that of the agonist. If an agonist is contracting during a specific movement, the antagonist is stretched. During flexion of the elbow, the biceps' brachii muscle contracts and acts as an agonist, while the triceps' brachii muscle on the opposite side of the upper arm acts as an antagonist and stretches.

Skeletal Muscles

Skeletal muscles, or voluntary muscles, are attached to bones by tendons and are responsible for voluntary movement. The connecting tendons are made up of dense bands of connective tissue and have collagen fibers that firmly attach the muscle to the bone. Their fibers are actually woven into the coverings of the bone and muscle so that they can withstand pressure and tension. They usually work in opposing pairs like the biceps and triceps, for example.

Skeletal muscles are made of bundles of long fibers that are composed of cells with many nuclei due to their length. These fibers contain myofibrils, and myofibrils are made of alternating filaments. The thicker myosin filaments are in between the smaller actin filaments in a unit called a sarcomere, and the overlapping regions of these filaments give the muscle its characteristic striated, or striped, appearance. Actin filaments are attached to exterior Z lines, myosin filaments are attached to a central M line, and when a muscle is at rest, there is a gap between the Z line and the myosin filaments. Only when the muscle contracts and the actin filaments slide over the myosin filaments does the myosin reach the Z line, as illustrated in the picture below. This sliding-filament model of muscle contraction is dependent on myosin molecules forming and breaking cross-bridges with actin in order to pull the actin filaments closer to the M line.

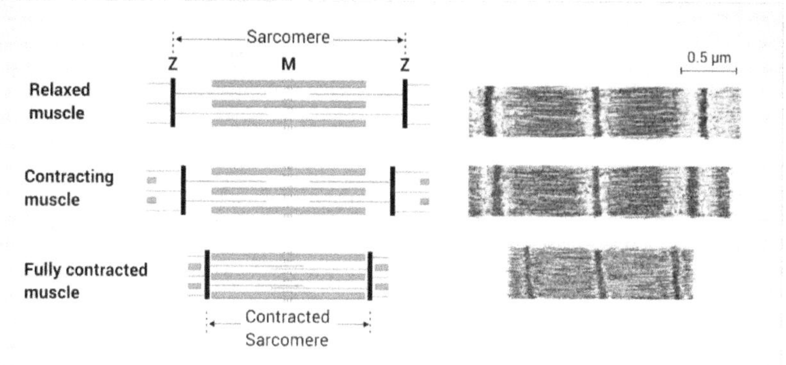

Skeletal muscles are controlled by the nervous system. Motor neurons connect to muscle fibers via neuromuscular junctions. Motor neurons must release the neurotransmitter acetylcholine which releases calcium ions to stimulate myosin cross-bridging and contraction. As the acetylcholine stops being released, the contraction ends.

Smooth Muscles

Smooth muscles are responsible for involuntary movement, such as moving food through the digestive tract and moving blood through vessels. Their cells have only one nucleus and do not have striations because actin and myosin filaments do not have an organized arrangement like they do in skeletal muscles. Unlike skeletal muscle, smooth muscle doesn't rely on neuromuscular junctions for intercellular communication. Instead, they operate via gap junctions, which send impulses directly from cell to cell.

Cardiac Muscles

Cardiac muscle cells are found only in the heart where they control the heart's rhythm and blood pressure. Like skeletal muscle, cardiac muscle has striations, but cardiac muscle cells are smaller than skeletal muscle cells, so they typically have only one nucleus. Like smooth muscle, cardiac muscles do not require neurotransmitter release by motor neurons to function, and they instead operate via gap junctions.

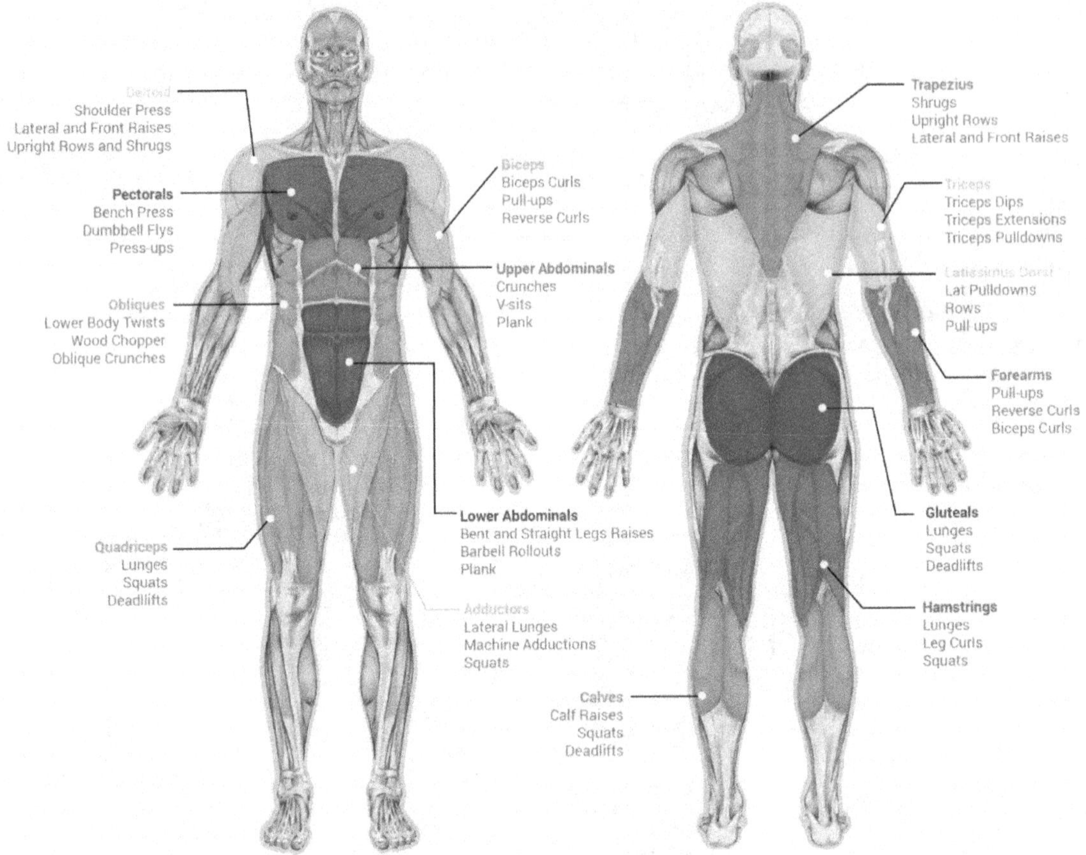

Nervous System

The human nervous system coordinates the body's response to stimuli from inside and outside the body. There are two major types of nervous system cells: neurons and neuroglia. Neurons are the workhorses of the nervous system and form a complex communication network that transmits electrical impulses termed action potentials, while neuroglia connect and support the neurons.

Although some neurons monitor the senses, some control muscles, and some connect the brain to other neurons, all neurons have four common characteristics:

- Dendrites: These receive electrical signals from other neurons across small gaps called *synapses*.
- Nerve cell body: This is the hub of processing and protein manufacture for the neuron.
- Axon: This transmits the signal from the cell body to other neurons.
- Terminals: These bridge the neuron to dendrites of other neurons and deliver the signal via chemical messengers called neurotransmitters.

Here is an illustration of this:

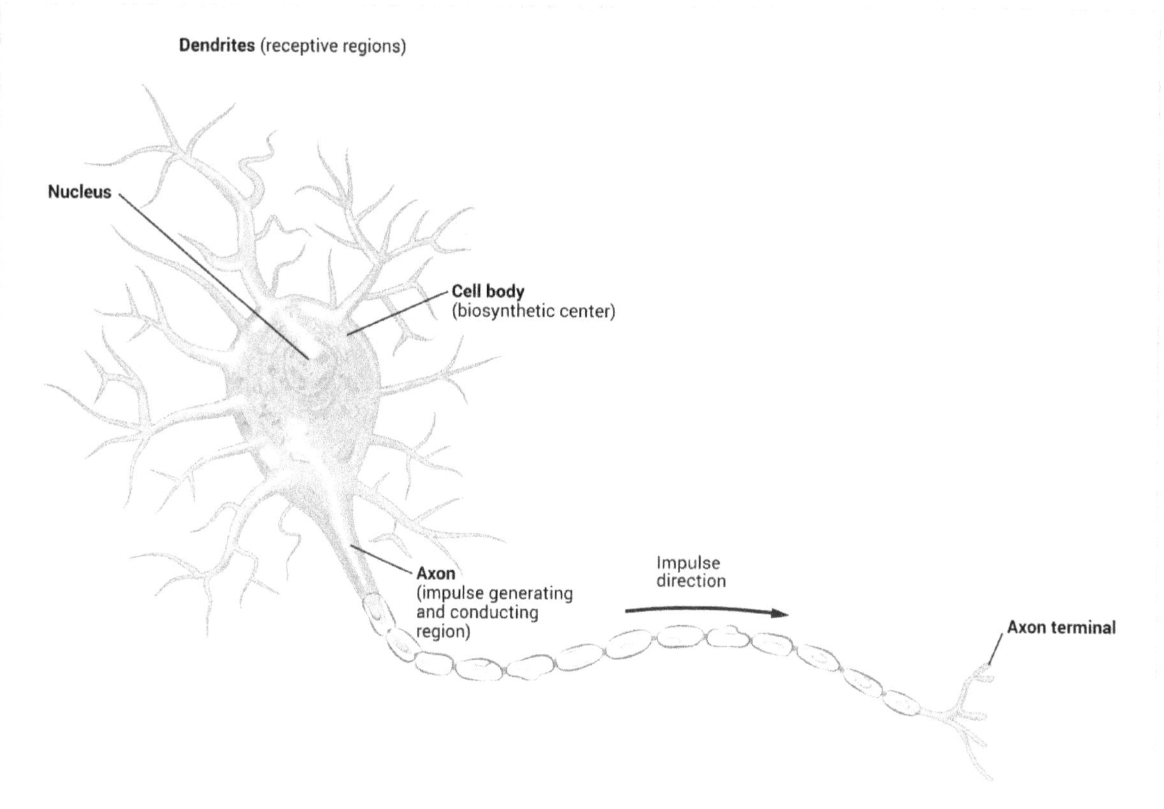

There are two major divisions of the nervous system, central and peripheral:

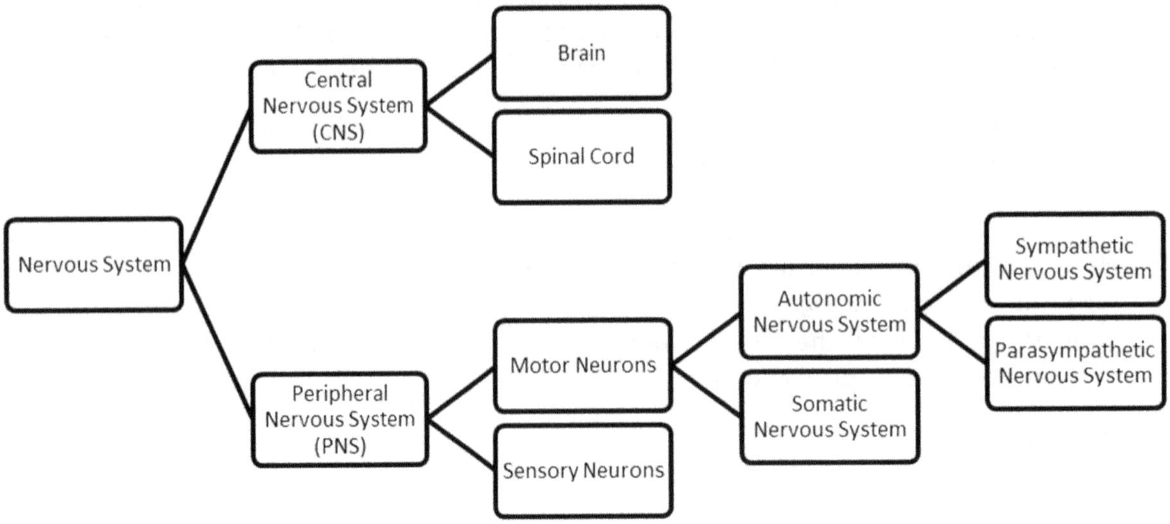

Central Nervous System

The central nervous system (CNS) consists of the brain and spinal cord. Three layers of membranes called the meninges cover and separate the CNS from the rest of the body.

The major divisions of the brain are the forebrain, the midbrain, and the hindbrain.

The forebrain consists of the cerebrum, the thalamus and hypothalamus, and the rest of the limbic system. The *cerebrum* is the largest part of the brain, and its most well-documented part is the outer cerebral cortex. The cerebrum is divided into right and left hemispheres, and each cerebral cortex hemisphere has four discrete areas, or lobes: frontal, temporal, parietal, and occipital. The frontal lobe governs duties such as voluntary movement, judgment, problem solving, and planning, while the other lobes are more sensory. The temporal lobe integrates hearing and language comprehension, the parietal lobe processes sensory input from the skin, and the occipital lobe processes visual input from the eyes. For completeness, the other two senses, smell and taste, are processed via the olfactory bulbs. The thalamus helps organize and coordinate all of this sensory input in a meaningful way for the brain to interpret.

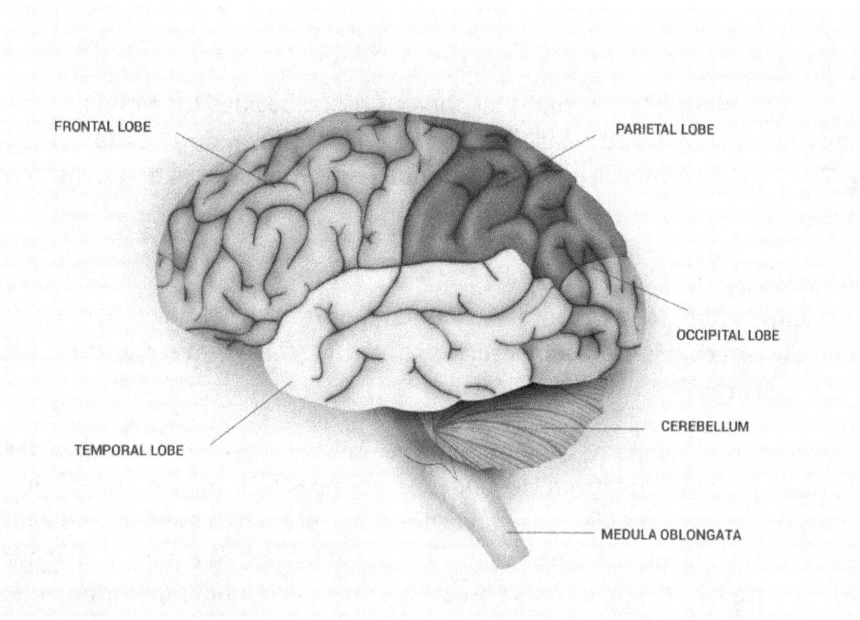

The hypothalamus controls the endocrine system and all of the hormones that govern long-term effects on the body. Each hemisphere of the limbic system includes a hippocampus (which plays a vital role in memory), an amygdala (which is involved with emotional responses like fear and anger), and other small bodies and nuclei associated with memory and pleasure.

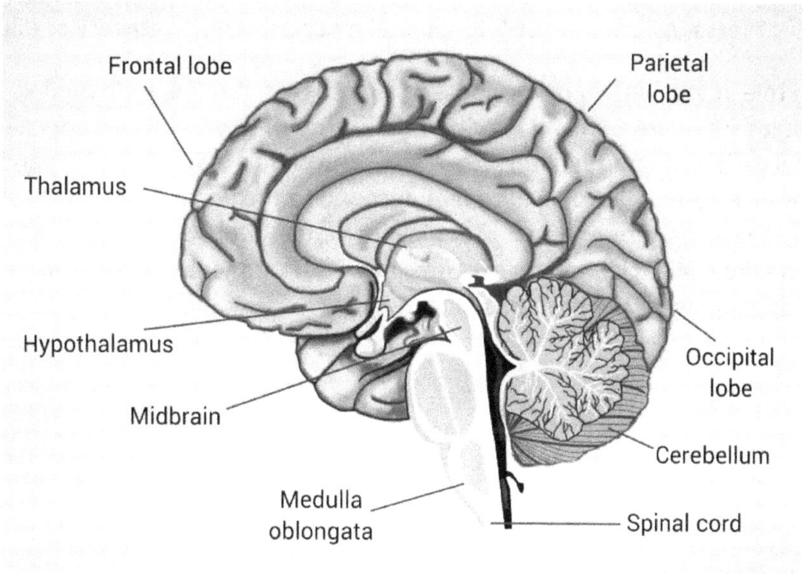

The midbrain is in charge of alertness, sleep/wake cycles, and temperature regulation, and it includes the substantia nigra which produces melatonin to regulate sleep patterns. The notable components of the hindbrain include the medulla oblongata and cerebellum. The **medulla oblongata** is located just above the spinal cord and is responsible for crucial involuntary functions such as breathing, swallowing, and regulating heart rate and blood pressure. Together with other parts of the hindbrain, the midbrain and medulla oblongata form the brain stem. The brain stem connects the spinal cord to the rest of the brain. To the rear of the brain stem sits the cerebellum, which plays key roles in posture, balance, and muscular coordination. The spinal cord itself, which is encapsulated by the protective bony spinal column, carries sensory information to the brain and motor information to the body.

Peripheral Nervous System

The peripheral nervous system (PNS) includes all nervous tissue besides the brain and spinal cord. The PNS consists of the sets of cranial and spinal nerves and relays information between the CNS and the rest of the body. The PNS has two divisions: the autonomic nervous system and the somatic nervous system.

Autonomic Nervous System

The autonomic nervous system (ANS) governs involuntary, or reflexive, body functions. Ultimately, the autonomic nervous system controls functions such as breathing, heart rate, digestion, body temperature, and blood pressure.

The ANS is split between parasympathetic nerves and sympathetic nerves. These two nerve types are antagonistic, and have opposite effects on the body. Parasympathetic nerves predominate resting conditions, and decrease heart rate, decrease breathing rate, prepare digestion, and allow urination and excretion. Sympathetic nerves, on the other hand, become active when a person is under stress or excited, and they increase heart rate, increase breathing rates, and inhibit digestion, urination, and excretion.

Somatic Nervous System and the Reflex Arc

The somatic nervous system (SNS) governs the conscious, or voluntary, control of skeletal muscles and their corresponding body movements. The SNS contains afferent and efferent neurons. Afferent neurons carry sensory messages from the skeletal muscles, skin, or sensory organs to the CNS. Efferent neurons relay motor messages from the CNS to skeletal muscles, skin, or sensory organs.

The SNS also has a role in involuntary movements called reflexes. A reflex is defined as an involuntary response to a stimulus. They are transmitted via what is termed a *reflex arc*, where a stimulus is sensed by an affector and its afferent neuron, interpreted and rerouted by an interneuron, and delivered to effector muscles by an efferent neuron where they respond to the initial stimulus. A reflex is able to bypass the brain by being rerouted through the spinal cord; the interneuron decides the proper course of action rather than the brain. The reflex arc results in an instantaneous, involuntary response. For example, a physician tapping on the knee produces an involuntary knee jerk referred to as the patellar tendon reflex.

Special Senses

The special senses include vision, hearing and balance, smell, and taste. They are distinguished from general senses in that special senses have *special somatic afferents* and *special visceral afferents*, both a type of nerve fiber relaying information to the CNS, as well as specialized organs devoted to their function. Touch is the other sense that is typically discussed, but unlike the special senses, it relays information to the CNS from all over the body and not just one particular organ; **skin**, the largest organ of the body, is the largest contributor to tactile information, but touch receptors also include **mechanoreceptors**, **nociceptors** for pain, and **thermoreceptors** for heat. Tactile messages are carried via *general somatic afferents* and *general visceral afferents.* Various touch receptors exist in the body such as the following:

- Pacinian corpuscles: detect rapid vibration in the skin and fascia
- Meissner's corpuscles: respond to light touch and slower vibrations
- Merkel's discs: respond to sustained pressure
- Ruffini endings: detect deep touch and tension in the skin and fascia

Endocrine System

The endocrine system is made up of the ductless tissues and glands that secrete hormones directly into the bloodstream. It is similar to the nervous system in that it controls various functions of the body, but it does so via secretion of hormones in the bloodstream as opposed to nerve impulses. The endocrine system is also different because its effects last longer than that of the nervous system. Nerve impulses are immediate while hormone responses can last for minutes or even days.

The endocrine system works closely with the nervous system to regulate the physiological activities of the other systems of the body in order to maintain homeostasis. Hormone secretions are controlled by tight feedback loops that are generally regulated by the hypothalamus, the bridge between the nervous and endocrine systems. The hypothalamus receives sensory input via the nervous system and responds by stimulating or inhibiting the pituitary gland, which, in turn, stimulates or inhibits several other glands. The tight control is due to hormone secretions.

Hormones are chemicals that bind to specific target cells. Each hormone will only bind to a target cell that has a specific receptor that has the correct shape. For example, testosterone will not attach to skin cells because skin cells have no receptor that recognizes testosterone.

There are two types of hormones: steroid and protein. Steroid hormones are lipid, nonpolar substances, and most are able to diffuse across cell membranes. Once they do, they bind to a receptor that initiates a signal transduction cascade that affects gene expression. Non-steroid hormones bind to receptors on cell membranes that also initiate a signal transduction cascade that affects enzyme activity and chemical reactions.

Major Endocrine Glands

Hypothalamus: This gland is a part of the brain. It connects the nervous system to the endocrine system because it receives sensory information through nerves, and it sends instructions via hormones delivered to the pituitary.

Pituitary Gland: This gland is pea-sized and is found at the bottom of the hypothalamus. It has two lobes called the anterior and posterior lobes. It plays an important role in regulating other endocrine glands. For example, it secretes growth hormone, which regulates growth. Other hormones that are released by this gland control the reproductive system, childbirth, nursing, blood osmolarity, and metabolism.

The hormones and glands respond to each other via feedback loops, and a typical feedback loop is illustrated in the picture below. The hypothalamus and pituitary gland are master controllers of most of the other glands.

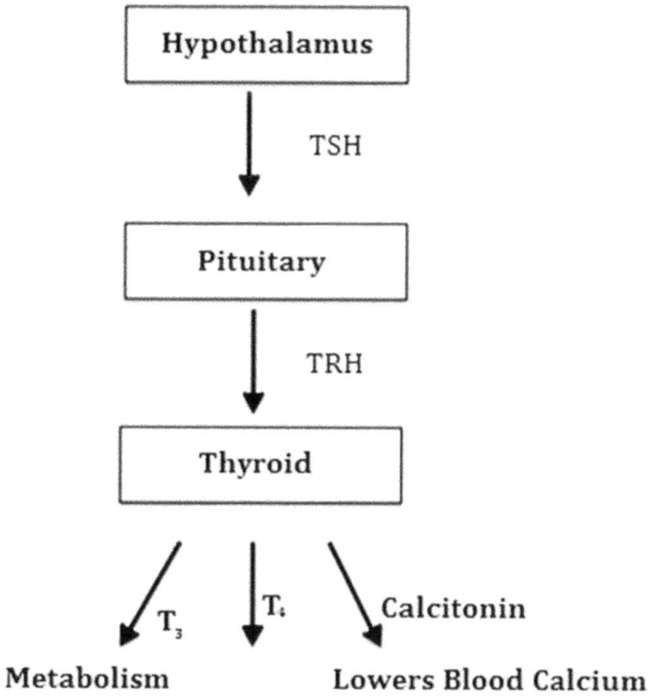

Thymus Gland: This gland is located in the chest cavity, embedded in connective tissue. It produces several hormones that are important for development and maintenance of T lymphocytes, which are important cells for immunity.

Adrenal Gland: One adrenal gland is attached to the top of each kidney. It produces epinephrine and norepinephrine which cause the "fight or flight" response in the face of danger or stress. These hormones raise heart rate, blood pressure, dilate bronchioles, and deliver blood to the muscles. All of these actions increase circulation and release glucose so that the body has an energy burst.

Pineal Gland: The pineal gland secretes melatonin, which is a hormone that regulates the body's circadian rhythm, which governs the natural wake-sleep cycle.

Testes and Ovaries: They secrete testosterone and both estrogen and progesterone, respectively. They are responsible for secondary sex characteristics, gamete development, and female hormones are important for embryonic development.

Thyroid Gland: This gland releases hormones like thyroxine and calcitonin. Thyroxine stimulates metabolism, and calcitonin monitors the amount of circulating calcium. Calcitonin signals the body to regulating calcium from bone reserves as well as kidney reabsorption of calcium.

Parathyroid Glands: These are four pea-sized glands located on the posterior surface of the thyroid. The main hormone that is secreted is called parathyroid hormone (PTH), which influences calcium levels like calcitonin, except it is antagonistic. PTH increases extracellular levels of calcium while calcitonin decreases it.

Pancreas: The pancreas is an organ that has both endocrine and exocrine functions. It functions outside of a typical feedback loop in that blood sugar seems to signal the pancreas itself. The endocrine functions are controlled by the pancreatic islets of Langerhans, which are groups of beta cells scattered throughout the gland that secrete insulin to lower blood sugar levels in the body. Neighboring alpha cells secrete glucagon to raise blood sugar. These complementary hormones keep blood sugar in check.

Cardiovascular System

The cardiovascular system (also called the circulatory system) is a network of organs and tubes that transport blood, hormones, nutrients, oxygen, and other gases to cells and tissues throughout the body. It is also known as the cardiovascular system. The major components of the circulatory system are the blood vessels, blood, and heart.

Blood Vessels

In the circulatory system, blood vessels are responsible for transporting blood throughout the body. The three major types of blood vessels in the circulatory system are arteries, veins, and capillaries. Arteries carry blood from the heart to the rest of the body. Veins carry blood from the body back to the heart. Capillaries connect arteries to veins and form networks that exchange materials between the blood and the cells.

In general, arteries are stronger and thicker than veins, as they withstand high pressures exerted by the blood as the heart pumps it through the body. Arteries control blood flow through either vasoconstriction (narrowing of the blood vessel's diameter) or vasodilation (widening of the blood vessel's diameter). The blood in veins is under much lower pressures, so veins have valves to prevent the backflow of blood.

Most of the exchange between the blood and tissues takes place through the capillaries. There are three types of capillaries: continuous, fenestrated, and sinusoidal.

Continuous capillaries are made up of epithelial cells tightly connected together. As a result, they limit the types of materials that pass into and out of the blood. Continuous capillaries are the most common type of capillary. Fenestrated capillaries have openings that allow materials to be freely exchanged between the blood and tissues. They are commonly found in the digestive, endocrine, and urinary systems. Sinusoidal capillaries have larger openings and allow proteins and blood cells through. They are found primarily in the liver, bone marrow, and spleen.

Blood

Blood is vital to the human body. It is a liquid connective tissue that serves as a transport system for supplying cells with nutrients and carrying away their wastes. The average adult human has five to six quarts of blood circulating through their body. Approximately 55% of blood is plasma (the fluid portion), and the remaining 45% is composed of solid cells and cell parts. There are three major types of blood cells:

- Red blood cells, or erythrocytes, transport oxygen throughout the body. They contain a protein called hemoglobin that allows them to carry oxygen. The iron in the hemoglobin gives the cells and the blood their red colors.

- White blood cells, or leukocytes, are responsible for fighting infectious diseases and maintaining the immune system. There are five types of white blood cells: neutrophils, lymphocytes, eosinophils, monocytes, and basophils.

- Platelets are cell fragments that play a central role in the blood clotting process.

All blood cells in adults are produced in the bone marrow—red blood cells and most white blood cells are produced in the red marrow, and some white blood cells are produced in the yellow bone marrow.

Heart

The heart is a two-part, muscular pump that forcefully pushes blood throughout the human body. The human heart has four chambers—two upper atria and two lower ventricles separated by a partition called the septum. There is a pair on the left and a pair on the right. Anatomically, *left* and *right* correspond to the sides of the body that the patient themselves would refer to as left and right.

Four valves help to section off the chambers from one another. Between the right atrium and ventricle, the three flaps of the tricuspid valve keep blood from backflowing from the ventricle to the atrium, similar to how the two flaps of the mitral valve work between the left atrium and ventricle. As these two valves lie between an atrium and a ventricle, they are referred to as atrioventricular (AV) valves. The other two valves are semilunar (SL) and control blood flow into the two great arteries leaving the ventricles. The pulmonary valve connects the right ventricle to the pulmonary artery, while the aortic valve connects the left ventricle to the aorta.

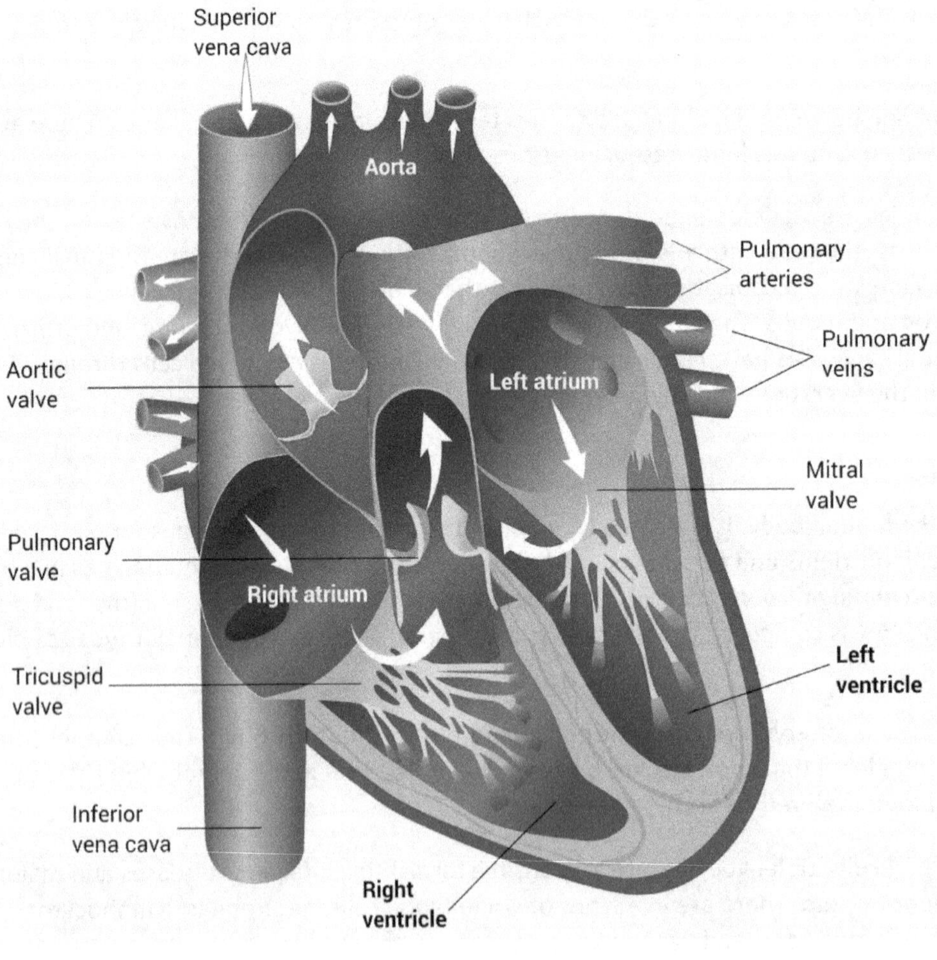

Cardiac Cycle

A cardiac cycle is one complete sequence of cardiac activity. The cardiac cycle represents the relaxation and contraction of the heart and can be divided into two phases: diastole and systole.

Diastole is the phase during which the heart relaxes and fills with blood. It gives rise to the diastolic blood pressure (DBP), which is the bottom number of a blood pressure reading. Systole is the phase during which the heart contracts and discharges blood. It gives rise to the systolic blood pressure (SBP), which is the top number of a blood pressure reading. The heart's electrical conduction system coordinates the cardiac cycle.

Types of Circulation

Five major blood vessels manage blood flow to and from the heart: the superior and inferior venae cava, the aorta, the pulmonary artery, and the pulmonary vein.

The superior vena cava is a large vein that drains blood from the head and the upper body. The inferior vena cava is a large vein that drains blood from the lower body. The aorta is the largest artery in the human body and carries blood from the heart to body tissues. The pulmonary arteries carry blood from the heart to the lungs. The pulmonary veins transport blood from the lungs to the heart.

In the human body, there are two types of circulation: pulmonary circulation and systemic circulation. Pulmonary circulation supplies blood to the lungs. Deoxygenated blood enters the right atrium of the heart and is routed through the tricuspid valve into the right ventricle. Deoxygenated blood then travels from the right ventricle of the heart through the pulmonary valve and into the pulmonary arteries. The pulmonary arteries carry the deoxygenated blood to the lungs. In the lungs, oxygen is absorbed, and carbon dioxide is released. The pulmonary veins carry oxygenated blood to the left atrium of the heart.

Systemic circulation supplies blood to all other parts of the body, except the lungs. Oxygenated blood flows from the left atrium of the heart through the mitral, or bicuspid, valve into the left ventricle of the heart. Oxygenated blood is then routed from the left ventricle of the heart through the aortic valve and into the aorta. The aorta delivers blood to the systemic arteries, which supply the body tissues. In the tissues, oxygen and nutrients are exchanged for carbon dioxide and other wastes. The deoxygenated blood along with carbon dioxide and wastes enter the systemic veins, where they are returned to the right atrium of the heart via the superior and inferior vena cava.

Lymphatic System and Immunity

The immune system is the body's defense against invading microorganisms (bacteria, viruses, fungi, and parasites) and other harmful, foreign substances. It is capable of limiting or preventing infection.

There are two general types of immunity: innate immunity and acquired immunity. Innate immunity uses physical and chemical barriers to block microorganism entry into the body. The biggest barrier is the skin; it forms a physical barrier that blocks microorganisms from entering underlying tissues. Mucous membranes in the digestive, respiratory, and urinary systems secrete mucus to block and remove invading microorganisms. Other natural defenses include saliva, tears, and stomach acids, which are all chemical barriers intended to block infection with microorganisms. Acid is inhospitable to pathogens, as are tears, mucus, and saliva, which all contain a natural antibiotic called lysozyme. The respiratory passages contain microscopic cilia, which are like bristles that sweep out pathogens. In addition, macrophages and other white blood cells can recognize and eliminate foreign objects through phagocytosis or toxic secretions.

Acquired immunity refers to a specific set of events used by the body to fight a particular infection. Essentially, the body accumulates and stores information about the nature of an invading microorganism. As a result, the body can mount a specific attack that is much more effective than innate immunity. It also provides a way for the body to prevent future infections by the same microorganism.

Acquired immunity is divided into a primary response and a secondary response. The primary immune response occurs the first time a particular microorganism enters the body, where macrophages engulf the microorganism and travel to the lymph nodes. In the lymph nodes, macrophages present the invader to helper T lymphocytes, which then activate humoral and cellular immunity. Humoral immunity refers to immunity resulting from antibody production by B lymphocytes. After being activated by helper T lymphocytes, B lymphocytes multiply and divide into plasma cells and memory cells. Plasma cells are B lymphocytes that produce immune proteins called antibodies, or immunoglobulins. Antibodies then bind the microorganism to flag it for destruction by other white blood cells. Cellular immunity refers to the immune response coordinated by T lymphocytes. After being activated by helper T lymphocytes, other T lymphocytes attack and kill cells that cause infection or disease.

The secondary immune response takes place during subsequent encounters with a known microorganism. Memory cells respond to the previously encountered microorganism by immediately producing antibodies. Memory cells are B lymphocytes that store information to produce antibodies. The secondary immune response is swift and powerful because it eliminates the need for the time-consuming macrophage activation of the primary immune response. Suppressor T lymphocytes also take part to inhibit the immune response as an overactive immune response could cause damage to healthy cells.

Inflammation occurs if a pathogen evades the barriers and chemical defenses. It stimulates pain receptors, alerting the individual that something is wrong. It also elevates body temperature to speed up chemical reactions, although if a fever goes unchecked it can be dangerous due to the fact that extreme heat unfolds proteins. Histamine is secreted which dilates blood vessels and recruits white blood cells that destroy invaders non-specifically. The immune system is tied to the lymphatic system. The thymus, one of the lymphatic system organs, is the site of maturation of T-cells, a type of white blood cell. The lymphatic system is important in the inflammatory response because lymph vessels deliver leukocytes and collect debris that will be filtered in the lymph nodes and the spleen.

Antigen and Typical Immune Response

Should a pathogen evade barriers and survive through inflammation, an antigen-specific adaptive immune response will begin. Immune cells recognize these foreign particles by their antigens, which are their unique and identifying surface proteins. Drugs, toxins, and transplanted cells can also act as antigens. The body even recognizes its own cells as potential threats in autoimmune diseases.

When a macrophage engulfs a pathogen and presents its antigens, helper T cells recognize the signal and secrete cytokines to signal T lymphocytes and B lymphocytes so that they launch the cell-mediated and humoral response, respectively. The cell-mediated response occurs when the T lymphocytes kill infected cells by secreting cytotoxins. The humoral response occurs when B lymphocytes proliferate into plasma and memory cells. The plasma cells secrete antigen-specific antibodies, which bind to the pathogens so that they cannot bind to host cells. Macrophages and other phagocytic cells called neutrophils engulf and degrade the antibody/pathogen complex. The memory cells remain in circulation and initiate a secondary immune response should the pathogen dare enter the host again.

Active and Passive Immunity

Acquired immunity occurs after the first antigen encounter. The first time the body mounts this immune response is called the primary immune response. Because the memory B cells store information about the antigen's structure, any subsequent immune response causes a secondary immune response, which is much faster and produces substantially more antibodies due to the presence of memory B cells. If the secondary immune response is strong and fast enough, it will fight off the pathogen before an individual becomes symptomatic. This is a natural means of acquiring immunity.

Vaccination is the process of inducing immunity. *Active immunization* refers to immunity gained by exposure to infectious microorganisms or viruses and can be *natural* or *artificial*. Natural immunization refers to an individual being exposed to an infectious organism as a part of daily life. For example, it was once common for parents to expose their children to childhood diseases such as measles or chicken pox. Artificial immunization refers to therapeutic exposure to an infectious organism as a way of protecting an individual from disease. Today, the medical community relies on artificial immunization as a way to induce immunity.

Vaccines are used for the development of active immunity. A vaccine contains a killed, weakened, or inactivated microorganism or virus that is administered through injection, by mouth, or by aerosol. Vaccinations are administered to prevent an infectious disease but do not always guarantee immunity. Due to circulating memory B cells after administration, the secondary response will fight off the pathogen should it be encountered again in many cases. Both illness and vaccinations cause active immunity.

Passive immunity refers to immunity gained by the introduction of antibodies. This introduction can be also be natural or artificial. The process occurs when antibodies from the mother's bloodstream are passed on to the bloodstream of the developing fetus. Breast milk can also transmit antibodies to a baby. Babies are born with passive immunity, which provides protection against general infection for approximately the first six months of its life.

Types of Leukocytes

There are many leukocytes, or white blood cells, involved in both innate and adaptive immunity. All are developed in bone marrow. Many have been mentioned in the text above, but a comprehensive list is included here for reference.

- Monocytes are large phagocytic cells.
    - Macrophages engulf pathogens and present their antigen. Some circulate, but others reside in lymphatic organs like the spleen and lymph nodes.
    - Dendritic cells are also phagocytic and antigen-presenting.
- Granulocytes are cells that contain secretory granules.
    - Neutrophils are the most abundant white blood cell. They are circulating and aggressive phagocytic cells that are part of innate immunity. They also secrete substances that are toxic to pathogens.
    - Basophils and mast cells secrete histamine, which stimulates the inflammatory response.
    - Eosinophils are found underneath mucous membranes and defend against multi-cellular parasites like worms. They have low phagocytic activity and primarily secrete destructive enzymes.
- T lymphocytes mature in the thymus.
    - Helper T cells recognize antigens presented by macrophages and dendritic cells and secrete cytokines, which mount the humoral and cell-mediated immune response.
    - Killer T cells are cytotoxic cells involved in the cell-mediated response by recognizing and poisoning infected cells.
    - Suppressor T cells suppress the adaptive immune response when there is no threat to conserve resources and energy.
    - Memory T cells remain in circulation to aid in the secondary immune response.
- B lymphocytes mature in bone marrow.
    - Plasma B cells secrete antigen-specific antibodies when signaled by Helper T cells and are degraded after the immune response.
    - Memory B cells store antigen-specific antibody making instructions and remain circulating after the immune response is over.
- Natural killer cells are part of innate immunity and patrol and identify suspect-material. They respond by secreting cytotoxic substances.

Respiratory System

The respiratory system enables breathing and supports the energy-making process in our cells. The respiratory system transports an essential reactant, oxygen, to cells so that they can produce energy in their mitochondria via cellular respiration. The respiratory system also removes carbon dioxide, a waste product of cellular respiration.

This system is divided into the upper respiratory system and the lower respiratory system. The upper system comprises the nose, the nasal cavity and sinuses, and the pharynx. The lower respiratory system comprises the larynx (voice box), the trachea (windpipe), the small passageways leading to the lungs, and the lungs.

The pathway of oxygen to the bloodstream begins with the nose and the mouth. Upon inhalation, air enters the nose and mouth and passes into the sinuses where it gets warmed, filtered, and humidified. The throat, or the pharynx, allows the entry of both food and air; however, only air moves into the trachea, or windpipe, since the epiglottis covers the trachea during swallowing and prevents food from entering. The trachea contains mucus and cilia. The mucus traps many airborne pathogens while the cilia act as bristles that sweep the pathogens away toward the top of the trachea where they are either swallowed or coughed out.

The top of the trachea itself has two vocal cords, which make up the larynx. At its bottom, the trachea forks into two major bronchi—one for each lung. These bronchi continue to branch into smaller and smaller bronchioles before terminating in grape-like air sacs called alveoli; these alveoli are surrounded by capillaries and provide the body with an enormous amount of surface area to exchange oxygen and carbon dioxide gases, in a process called external respiration.

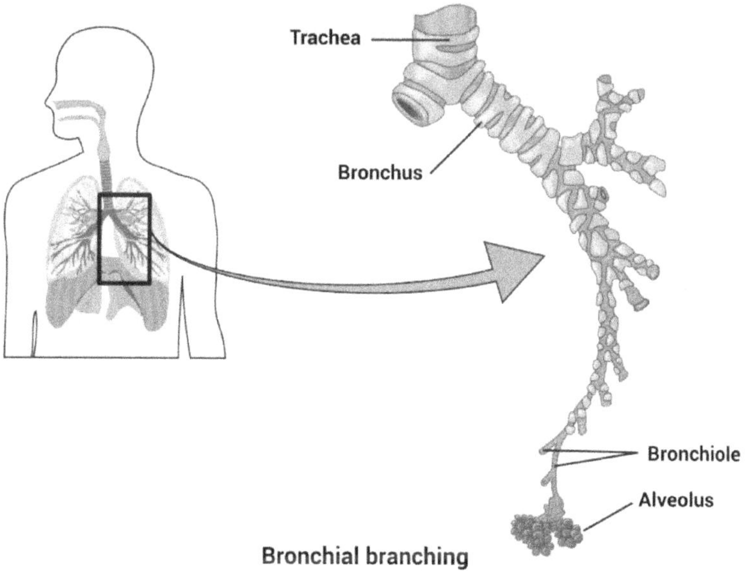

Bronchial branching

In total, the lungs contain about 1500 miles of airway passages. The right lung is divided into three lobes (superior, middle, and inferior), and the left lung is divided into two lobes (superior and inferior).

The left lung is smaller than the right lung, likely because it shares its space in the chest cavity with the heart.

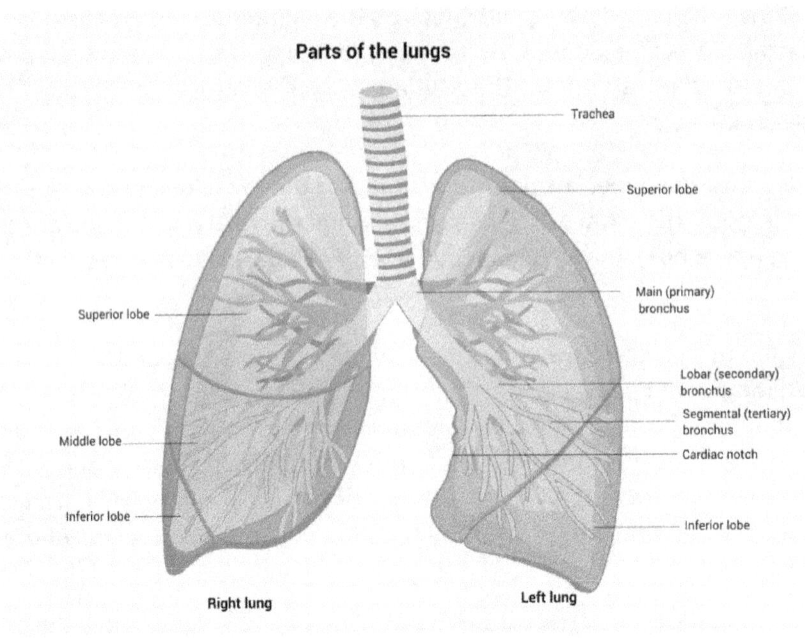

A flat muscle underneath the lungs called the diaphragm controls breathing. When the diaphragm contracts, the volume of the chest cavity increases and indirectly decreases its air pressure. This decrease in air pressure creates a vacuum, and the lungs pull in air to fill the space. This difference in air pressure pulls the air from outside of the body into the lungs in a process called negative pressure breathing.

Upon inhalation or inspiration, oxygen in the alveoli diffuses into the capillaries to be carried by blood to cells throughout the body, in a process called internal respiration. A protein called hemoglobin in red blood cells easily bonds with oxygen, removing it from the blood and allowing more oxygen to diffuse in. This protein allows the blood to take in 60 times more oxygen than the body could without it, and this explains how oxygen can become so concentrated in blood even though it is only 21% of the atmosphere. While oxygen diffuses from the alveoli into the capillaries, carbon dioxide diffuses from the capillaries into the alveoli. When the diaphragm relaxes, the elastic lungs snap back to their original shape; this decreases the volume of the chest cavity and increases the air pressure until it is back to normal. This increased air pressure pushes the carbon dioxide waste from the alveoli through exhalation or *expiration*.

The autonomic nervous system controls breathing. The medulla oblongata gets feedback regarding the carbon dioxide levels in the blood and will send a message to the diaphragm that it is time for a contraction. While breathing can be voluntary, it is mostly under autonomic control.

Functions of the Respiratory System

The respiratory system has many functions. Most importantly, it provides a large area for gas exchange between the air and the circulating blood. It protects the delicate respiratory surfaces from environmental variations and defends them against pathogens. It is responsible for producing the sounds that the body makes for speaking and singing, as well as for non-verbal communication. It also helps regulate blood volume and blood pressure by releasing vasopressin, and it is a regulator of blood pH due to its control over carbon dioxide release, as the aqueous form of carbon dioxide is the chief buffering agent in blood.

Digestive System

The human body relies completely on the digestive system to meet its nutritional needs. After food and drink are ingested, the digestive system breaks them down into their component nutrients and absorbs them so that the circulatory system can transport them to other cells to use for growth, energy, and cell repair. These nutrients may be classified as proteins, lipids, carbohydrates, vitamins, and minerals.

The digestive system is thought of chiefly in two parts: the digestive tract (also called the alimentary tract or gastrointestinal tract) and the accessory digestive organs. The digestive tract is the pathway in which food is ingested, digested, absorbed, and excreted. It is composed of the mouth, pharynx, esophagus, stomach, small and large intestines, rectum, and anus. *Peristalsis*, or wave-like contractions of smooth muscle, moves food and wastes through the digestive tract. The accessory digestive organs are the salivary glands, liver, gallbladder, and pancreas.

Mouth and Stomach

The mouth is the entrance to the digestive system. Here, the mechanical and chemical digestion of the food begins. The food is chewed mechanically by the teeth and shaped into a *bolus* by the tongue so that it can be more easily swallowed by the esophagus. The food also becomes more watery and pliable with the addition of saliva secreted from the salivary glands, the largest of which are the parotid glands. The glands also secrete amylase in the saliva, an enzyme which begins chemical digestion and breakdown of the carbohydrates and sugars in the food.

The food then moves through the pharynx and down the muscular esophagus to the stomach.

The stomach is a large, muscular sac-like organ at the distal end of the esophagus. Here, the bolus is subjected to more mechanical and chemical digestion. As it passes through the stomach, it is physically squeezed and crushed while additional secretions turn it into a watery nutrient-filled liquid that exits into the small intestine as *chyme*.

The stomach secretes a great many substances into the *lumen* of the digestive tract. Some cells produce gastrin, a hormone that prompts other cells in the stomach to secrete a gastric acid composed mostly of hydrochloric acid (HCl). The HCl is at such a high concentration and low pH that it denatures most proteins and degrades a lot of organic matter. The stomach also secretes mucous to form a protective film that keeps the corrosive acid from dissolving its own cells. Gaps in this mucous layer can lead to peptic ulcers. Finally, the stomach also uses digestive enzymes like proteases and lipases to break down proteins and fats; although there are some gastric lipases here, the stomach is most known for breaking down proteins.

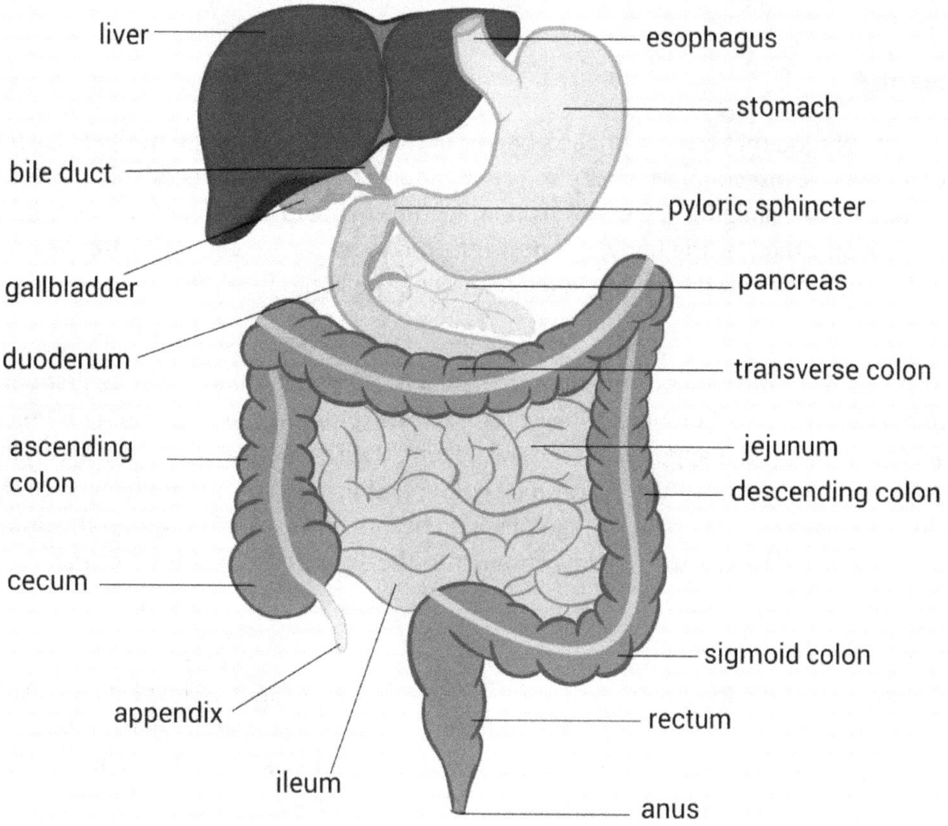

Small Intestine

The chyme from the stomach enters the first part of the small intestine, the *duodenum*, through the *pyloric sphincter*, and its extreme acidity is partly neutralized by sodium bicarbonate secreted along with mucous. The presence of chyme in the duodenum triggers the secretion of the hormones secretin and cholecystokinin (CCK). Secretin acts on the pancreas to dump more sodium bicarbonate into the small intestine so that the pH is kept at a reasonable level, while CCK acts on the gallbladder to release the *bile* that it has been storing. Bile, a substance produced by the liver and stored in the gallbladder, helps to emulsify or dissolve fats and lipids.

Because of the bile (which aids in lipid absorption) and the secreted lipases (which break down fats), the duodenum is the chief site of fat digestion in the body. The duodenum also represents the last major site of chemical digestion in the digestive tract, as the other two sections of the small intestine (the *jejunum* and *ileum*) are instead heavily involved in absorption of nutrients.

The small intestine reaches 40 feet in length, and its cells are arranged in small finger-like projections called villi. This is due to its key role in the absorption of nearly all nutrients from the ingested and digested food, effectively transferring them from the lumen of the GI tract to the bloodstream, where they travel to the cells that need them. These nutrients include simple sugars like glucose from carbohydrates, amino acids from proteins, emulsified fats, electrolytes like sodium and potassium, minerals like iron and zinc, and vitamins like D and B12. Vitamin B12's absorption, though it takes place in the intestines, is actually aided by *intrinsic factor* that was released into the chyme back in the stomach.

Large Intestine

The leftover parts of food that remain unabsorbed or undigested in the lumen of the small intestine next travel through the large intestine, which is also referred to as the large bowel or colon. The large intestine is mainly responsible for water absorption. As the chyme at this stage no longer has anything useful that can be absorbed by the body, it is now referred to as *waste*, and it is stored in the large intestine until it can be excreted from the body. Removing the liquid from the waste transforms it from liquid to solid stool, or feces.

This waste first passes from the small intestine to the cecum, a pouch that forms the first part of the large intestine. In herbivores, it provides a place for bacteria to digest cellulose, but in humans most of it is vestigial and is known as the appendix. From the cecum, waste next travels up the ascending colon, across the transverse colon, down the descending colon, and through the sigmoid colon to the rectum. The rectum is responsible for the final storage of waste before it is expelled through the anus. The anal canal is a small portion of the rectum leading through to the anus and the outside of the body.

Pancreas

The pancreas has endocrine and exocrine functions. The endocrine function involves releasing the hormone insulin, which decreases blood sugar (glucose) levels, and glucagon, which increases blood sugar (glucose) levels, directly into the bloodstream. Both hormones are produced in the islets of Langerhans, insulin in the beta cells and glucagon in the alpha cells.

The exocrine function of the pancreas involves the secretion of inactive digestive enzymes (zymogens) from acinar cells into the main pancreatic duct. The main pancreatic duct joins the common bile duct, which empties into the small intestine (specifically the duodenum). The digestive enzymes are then activated and take part in the digestion of carbohydrates, proteins, and fats within chyme (the mixture of partially digested food and digestive juices).

Metabolism

Metabolism is the set of chemical processes that occur within a cell for the maintenance of life. It includes both the synthesizing and breaking down of substances. A metabolic pathway begins with a molecule and ends with a specific product after going through a series of reactions, often involving an enzyme at each step. An enzyme is a protein that aids in the reaction. Catabolic pathways are metabolic pathways in which energy is released by complex molecules being broken down into simpler molecules. Contrast to catabolic pathways are anabolic pathways, which use energy to build complex molecules out of simple molecules. With cell metabolism, remember the first law of thermodynamics: Energy can be transformed, but it cannot be created or destroyed. Therefore, the energy released in a cell by a catabolic pathway is used up in anabolic pathways.

The reactions that occur within metabolic pathways are classified as either exergonic reactions or endergonic reactions. Exergonic reactions end in a release of free energy, while endergonic reactions absorb free energy from its surroundings. Free energy is the portion of energy in a system, such as a living cell, that can be used to perform work, such as a chemical reaction. It is denoted as the capital letter G and the change in free energy from a reaction or set of reactions is denoted as delta G (ΔG). When reactions do not require an input of energy, they are said to occur spontaneously. Exergonic reactions are considered spontaneous because they result in a negative delta G ($-\Delta G$), where the products of the reaction have less free energy within them than the reactants. Endergonic reactions require an input of energy and result in a positive delta G ($+\Delta G$), with the products of the reaction containing more free energy than the individual reactants. When a system no longer has free energy to do work, it has reached equilibrium. Since cells always work, they are no longer alive if they reach equilibrium.

Cells balance their energy resources by using the energy from exergonic reactions to drive endergonic reactions forward, a process called energy coupling. Adenosine triphosphate, or ATP, is a molecule that is an immediate source of energy for cellular work. When it is broken down, it releases energy used in endergonic reactions and anabolic pathways. ATP breaks down into adenosine diphosphate, or ADP, and a separate phosphate group, releasing energy in an exergonic reaction. As ATP is used up by reactions, it is also regenerated by having a new phosphate group added onto the ADP products within the cell in an endergonic reaction.

Enzymes are special proteins that help speed up metabolic reactions and pathways. They do not change the overall free energy release or consumption of reactions; they just make the reactions occur more quickly as it lowers the activation energy required. Enzymes are designed to act only on specific substrates. Their physical shape fits snugly onto their matched substrates, so enzymes only speed up reactions that contain the substrates to which they are matched.

Urinary System

The urinary system is made up of the kidneys, ureters, urinary bladder, and the urethra. It is the system responsible for removing waste products and balancing water and electrolyte concentrations in the blood. The urinary system has many important functions related to waste excretion. It regulates the concentrations of sodium, potassium, chloride, calcium, and other ions in the filtrate by controlling the amount of each that is reabsorbed during filtration. The reabsorption or secretion of hydrogen ions and bicarbonate contributes to the maintenance of blood pH. Certain kidney cells can detect any reductions in blood volume and pressure. If that happens, they secrete renin, which activates a hormone that causes increased reabsorption of sodium ions and water, raising volume and pressure. Under hypoxic conditions, kidney cells will secrete erythropoietin in order to stimulate red blood cell production. Kidney cells also synthesize calcitriol, which is a hormone derivative of vitamin D3 that aids in calcium ion absorption by the intestinal epithelium.

Under normal circumstances, humans have two functioning kidneys in the lower back and on either side of the spinal cord. They are the main organs that are responsible for filtering waste products out of the blood and regulating blood water and electrolyte levels. Blood enters the kidney through the renal artery, and urea and wastes are removed, while water and the acidity/alkalinity of the blood is adjusted. Toxic substances and drugs are also filtered. Blood exits through the renal vein, and the urine waste travels through the ureter to the bladder, where it is stored until it is eliminated through the urethra.

The kidneys have an outer renal cortex and an inner renal medulla that contain millions of tiny filtering units called nephrons. Nephrons have two parts: a glomerulus, which is the filter, and a tubule. The glomerulus is a network of capillaries covered by the Bowman's capsule, which is the entrance to the tubule. As blood enters the kidneys via the renal artery, the glomerulus allows for fluid and waste products to pass through it and enter the tubule. Blood cells and large molecules, such as proteins, do not pass through and remain in the blood.

The filtrate passes through the tubule, which has several parts. The proximal tubule comes first, and then the descending and ascending limbs of the loop of Henle dip into the medulla, followed by the distal tubule and collecting duct. The journey through the tubule involves a balancing act that regulates blood osmolarity, pH, and electrolytes exchange of materials between the tubule and the blood stream. The final product at the collecting tubule is called urine, and it is delivered to the bladder by the ureter. The most central part of the kidney is the renal pelvis, and it acts as a funnel by delivering the urine from the millions of the collecting tubules to the ureters. The filtered blood exits through the renal vein and is returned to circulation.

Here's a look at the genitourinary system:

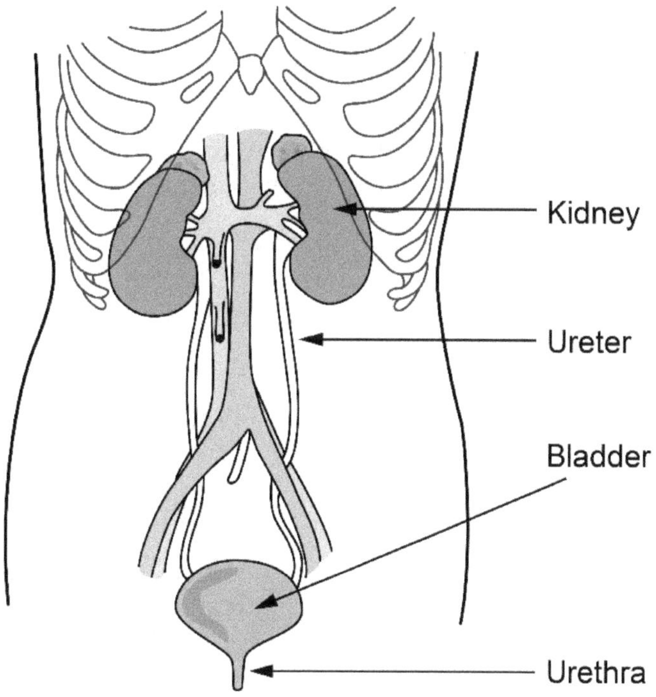

Here's a close up look at the kidney:

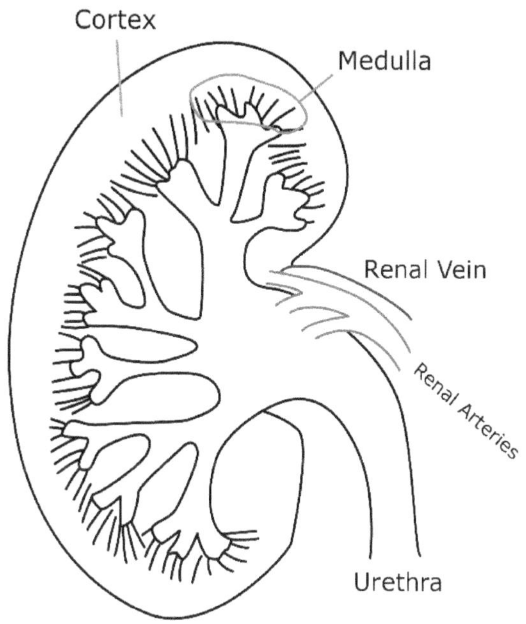

Waste Excretion

Once urine accumulates, it leaves the kidneys. The urine travels through the ureters into the urinary bladder, a muscular organ that is hollow and elastic. As more urine enters the urinary bladder, its walls stretch and become thinner so there is no significant difference in internal pressure. The urinary bladder stores the urine until the body is ready for urination, at which time the muscles contract and force the urine through the urethra and out of the body.

Fluid Electrolyte, Acid-Base Balance

Electrolytes are materials that conduct electricity due to ionization when dissolved in a solvent. Many acids and bases are electrolytes and most of their related salts are electrolytes as well. The *Brønsted-Lowry Acid-Base Theory* defines an acid as a substance that is a proton-donor and a base as a substance that accepts protons. The theory states that when an acid and base react with each other, the acid donates its proton and forms its *conjugate base*. Likewise, a base that has gained a proton becomes its *conjugate acid*. A neutralization of solution pH occurs when the acid and base are in equilibrium.

Water is the universal solvent and is amphoteric, which means that it can act as an acid or a base depending on the properties of the solute. In addition to H_2O molecules, water also contains a small concentration of dissociated H_3O^+ and OH^- ions. The strength of an aqueous acid or base depends on the extent of dissociation of the electrolyte in water. The more complete the ionization, the stronger the solution.

Reproductive System

The reproductive system is responsible for producing, storing, nourishing, and transporting functional reproductive cells, or gametes, in the human body. It includes the reproductive organs, also known as gonads, the reproductive tract, the accessory glands and organs that secrete fluids into the reproductive tract, and the perineal structures, which are the external genitalia.

Reproduction involves the passing of genes from one generation to the next, and that is accomplished through haploid gametes. Gametes have gone through meiosis and have 23 chromosomes, half the normal number. The male gamete is sperm, and the female gamete is an egg or ovum. When a sperm fertilizes an egg, they create a zygote, which is the first cell of a new organism. The zygote has a full set of 46 chromosomes because it received 23 from each parent. Because of gene shuffling during sperm and egg development, sperm and egg chromosome sets are all different, which results in the variety seen in humans.

Male Reproductive System

The entire male reproductive system is designed to generate sperm and produce semen that facilitate fertilization of eggs, the female gametes. The testes are the endocrine glands that secrete testosterone, a hormone that is important for secondary sex characteristics and sperm development, or spermatogenesis. Testosterone is in the androgen steroid-hormone family. The testes also produce and store 500 million spermatocytes, which are the male gametes, each day. Testes are housed in the scrotum, which is a sac that hangs outside the body so that spermatogenesis occurs at cooler and optimal conditions.

The seminiferous tubules within the testes produce spermatocytes, which then travel to the epididymis where they are stored as they mature. Then, the sperm move to the ejaculatory duct via the vas deferens. The ejaculatory duct contains more than just sperm. The seminal vesicles secrete an alkaline substance that will help sperm survive in the acidic vagina. The prostate gland secretes enzymes bathed in a milky white fluid that is important for thinning semen after ejaculation to increase its likelihood of reaching the egg. The bulbourethral gland, or Cowper's gland, secretes an alkaline fluid that lubricates the urethra prior to ejaculation to neutralize any acidic urine residue.

The sperm, along with all the exocrine secretions, are collectively called semen. In order for sexual intercourse to successfully transfer sperm from the male to the female's egg, the penis must be erect, which occurs via arousal and increased circulation. During sexual intercourse, ejaculation will forcefully expel the contents of the semen and effectively deliver the sperm to the egg. The muscular prostate gland is important for ejaculation. Each ejaculation releases 2 to 6 million sperm. Sperm has a whip-like flagellum tail that facilitates movement.

Female Reproductive System

The vagina is the passageway that sperm must travel through to reach an egg, the female gamete. Surrounding the vagina are the labia minor and labia major, both of which are folds that protect the urethra and the vaginal opening. The clitoris is rich in nerve-endings, making it sensitive and highly stimulated during sexual intercourse. It is above the vagina and urethra. An exocrine gland called the Bartholin's glands secretes a fluid during arousal that is important for lubrication.

The female gonads are the ovaries. Ovaries generally produce one immature gamete, an egg or oocyte, per month. They are also responsible for secreting the hormones estrogen and progesterone. Fertilization cannot happen unless the ejaculated sperm finds the egg, which is only available at certain times of the month. Eggs, or ova, develop in the ovaries in clusters surrounded by follicles, and after puberty, they are delivered to the uterus once a month via the fallopian tubes. The 28-day average journey of the egg to the uterus is called the menstrual cycle, and it is highly regulated by the endocrine system. The regulatory hormones Gonadotropin releasing hormone (GnRH), luteinizing hormone (LH), and follicle-stimulating hormone (FSH) orchestrate the menstrual cycle. Ovarian hormones estrogen and progesterone are also important in timing as well as for vascularization of the uterus in preparation for pregnancy. Fertilization usually happens around ovulation, which is when the egg is inside the fallopian tube. The resulting zygote travels down the tube and implants into the uterine wall. The uterus protects and nourishes the developing embryo for nine months until it is ready for the outside environment.

If the egg released is unfertilized, the uterine lining will slough off during menstruation. Should a fertilized egg, called a zygote, reach the uterus, it will embed itself into the uterine wall due to uterine vascularization that will deliver blood, nutrients, and antibodies to the developing embryo. The uterus is where the embryo will develop for the next nine months. Mammary glands are important female reproductive structures because they produce the milk provided for babies during lactation. Milk contains nutrients and antibodies that benefit the baby.

HAPS Practice Questions

1. The somatic nervous system is responsible for which of the following?
    a. Breathing
    b. Thought
    c. Movement
    d. Fear

2. The process of breaking large molecules into smaller molecules to provide energy is known as which of the following?
    a. Metabolism
    b. Bioenergetics
    c. Anabolism
    d. Catabolism

3. What type of vessel carries oxygen-rich blood from the heart to other tissues of the body?
    a. Veins
    b. Intestines
    c. Bronchioles
    d. Arteries

4. Which system comprises the 206 bones of the body?
    a. Skeletal
    b. Muscular
    c. Endocrine
    d. Reproductive

5. Which blood component is chiefly responsible for clotting?
    a. Platelets
    b. Red blood cells
    c. Antigens
    d. Plasma cells

6. Which is the first event to happen in a primary immune response?
    a. Macrophages ingest pathogens and present their antigens.
    b. Neutrophils aggregate and act as cytotoxic, nonspecific killers of pathogens.
    c. B lymphocytes make pathogen-specific antibodies.
    d. Helper T cells secrete interleukins to activate pathogen-fighting cells.

7. Where does sperm maturation take place in the male reproductive system?
    a. Seminal vesicles
    b. Prostate gland
    c. Epididymis
    d. Vas Deferens

8. Which of the following lists of joint types is in the correct order for increasing amounts of permitted motion (least mobile to most mobile)?
   a. Hinge, condyloid, saddle
   b. Saddle, hinge, condyloid
   c. Saddle, condyloid, hinge
   d. Hinge, saddle, condyloid

9. Which of the following is NOT a major function of the respiratory system in humans?
   a. It provides a large surface area for gas exchange of oxygen and carbon dioxide.
   b. It helps regulate the blood's pH.
   c. It helps cushion the heart against jarring motions.
   d. It is responsible for vocalization.

10. The function of synergists can best be described as which of the following?
    I. They assist primary movers in completing the specific movement
    II. They stabilize the point of origin and provide extra pull near the insertion
    III. They help prevent unwanted movement at a joint

   a. I, II
   b. I, III
   c. II, III
   d. All of the above

11. What makes bone resistant to shattering?
   a. The calcium salts deposited in the bone
   b. The collagen fibers
   c. The bone marrow and network of blood vessels
   d. The intricate balance of minerals and collagen fibers

12. Using anatomical terms, what is the relationship of the sternum relative to the deltoid?
   a. Medial
   b. Lateral
   c. Superficial
   d. Posterior

13. Which statement about white blood cells is true?
   a. B cells are responsible for antibody production.
   b. White blood cells are made in the white/yellow cartilage before they enter the bloodstream.
   c. Platelets, a special class of white blood cell, function to clot blood and stop bleeding.
   d. The majority of white blood cells only activate during the age of puberty, which explains why children and the elderly are particularly susceptible to disease.

14. Which locations in the digestive system are sites of chemical digestion?

I. Mouth
II. Stomach
III. Small Intestine

a. II only
b. III only
c. II and III only
d. I, II, and III

15. Which of the following structures is unique to eukaryotic cells?
a. Cell walls
b. Nuclei
c. Cell membranes
d. Organelles

16. Which is the cellular organelle used for digestion to recycle materials?
a. The Golgi apparatus
b. The lysosome
c. The centrioles
d. The mitochondria

17. Why do arteries have valves?
a. They have valves to maintain high blood pressure so that capillaries diffuse nutrients properly.
b. Their valves are designed to prevent backflow due to their low blood pressure.
c. The valves have no known purpose and thus appear to be unnecessary.
d. They do not have valves, but veins do.

18. If the pressure in the pulmonary artery is increased above normal, which chamber of the heart will be affected first?
a. The right atrium
b. The left atrium
c. The right ventricle
d. The left ventricle

19. What is the purpose of sodium bicarbonate when released into the lumen of the small intestine?
a. It works to chemically digest fats in the chyme.
b. It decreases the pH of the chyme so as to prevent harm to the intestine.
c. It works to chemically digest proteins in the chyme.
d. It increases the pH of the chyme so as to prevent harm to the intestine.

20. Which of the following describes a reflex arc?
a. The storage and recall of memory
b. The maintenance of visual and auditory acuity
c. The autoregulation of heart rate and blood pressure
d. A stimulus and response controlled by the spinal cord

21. Nociceptors detect which of the following?
   a. Deep pressure
   b. Vibration
   c. Pain
   d. Temperature

22. Ligaments connect what?
   a. Muscle to muscle
   b. Bone to bone
   c. Bone to muscle
   d. Muscle to tendon

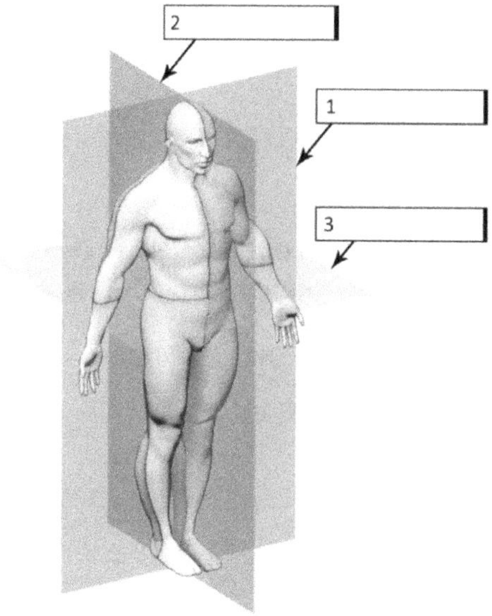

23. Identify the correct sequence of the 3 primary body planes as numbered 1, 2, and 3 in the above image.
   a. Plane 1 is coronal, plane 2 is sagittal, and plane 3 is transverse.
   b. Plane 1 is sagittal, plane 2 is coronal, and plane 3 is medial.
   c. Plane 1 is coronal, plane 2 is sagittal, and plane 3 is medial.
   d. Plane 1 is sagittal, plane 2 is coronal, and plane 3 is transverse.

24. The strength of an aqueous acid or base depends on which of the following?
   a. The extent to which its electrolytes dissociates in water
   b. The extent to which it acts as a universal solvent.
   c. The extent to which it can donate its proton and form its conjugate base
   d. The extent to which it can gain a proton and form its conjugate acid

25. Which of the following is NOT a function of the forebrain?
   a. To regulate blood pressure and heart rate
   b. To perceive and interpret emotional responses like fear and anger
   c. To perceive and interpret visual input from the eyes
   d. To integrate voluntary movement

26. A child complains of heavy breathing even when relaxing. They are an otherwise healthy child with no history of respiratory problems. What might be the issue?
   a. Asthma
   b. Blood clot
   c. Hyperventilation
   d. Exercising too hard

27. What coefficients are needed to balance the following combustion equation?

$$\_\,C_2H_{10} + \_\,O_2 \rightarrow \_\,H_2O + \_\,CO_2$$

   a. 1:5:5:2
   b. 1:9:5:2
   c. 2:9:10:4
   d. 2:5:10:4

28. Which statement is true about the pH of a solution?
   a. A solution cannot have a pH less than 1.
   b. The more hydroxide ions in the solution, the higher the pH.
   c. If an acid has a pH of greater than 2, it is considered a weak base.
   d. A solution with a pH of 2 has ten times more hydrogen ions than a solution with a pH of 1.

29. Salts like sodium iodide (NaI) and potassium chloride (KCl) use what type of bond?
   a. Ionic bonds
   b. Disulfide bridges
   c. Covalent bonds
   d. London dispersion forces

30. Which of the following is unique to covalent bonds?
   a. Most covalent bonds are formed between the elements H, F, N, and O.
   b. Covalent bonds are dependent on forming dipoles.
   c. Bonding electrons are shared between two or more atoms.
   d. Molecules with covalent bonds tend to have a crystalline solid structure.

31. Which of the following areas of the body has the most sweat glands?
   a. Upper back
   b. Arms
   c. Feet
   d. Palms

32. A patient's body is not properly filtering blood. Which of the following body parts is most likely malfunctioning?
   a. Medulla
   B. Heart
   C. Nephrons
   D. Renal cortex

33. A pediatrician notes that an infant's cartilage is disappearing and being replaced by bone. What process has the doctor observed?
   a. Mineralization

b. Ossification
c. Osteoporosis
d. Calcification

34. Which of the following is a special property of water?
a. Water easily flows through phospholipid bilayers.
b. A water molecule's oxygen atom allows fish to breathe.
c. Water is highly cohesive which explains its high melting point.
d. Water can self-hydrolyze and decompose into hydrogen and oxygen.

35. What is the electrical charge of the nucleus?
a. A nucleus always has a positive charge.
b. A stable nucleus has a positive charge, but a radioactive nucleus may have no charge and instead be neutral.
c. A nucleus always has no charge and is instead neutral.
d. A stable nucleus has no charge and is instead neutral, but a radioactive nucleus may have a charge.

36. Which of the following creates sperm?
a. Prostate gland
b. Seminal vesicles
c. Scrotum
d. Seminiferous tubules

37. Which of the following functions corresponds to the parasympathetic nervous system?
a. It stimulates the fight-or-flight response.
b. It increases heart rate.
c. It stimulates digestion.
d. It increases bronchiole dilation.

38. Which of the following is the gland that helps regulate calcium levels?
a. Osteotoid gland
b. Pineal gland
c. Parathyroid glands
d. Thymus gland

39. Which of the following systems does not include a transportation system throughout the body?
a. Cardiovascular system
b. Endocrine system
c. Immune system
d. Nervous system

40. Which of the following correctly identifies a difference between the primary and secondary immune response?
a. In the secondary response, macrophages migrate to the lymph nodes to present the foreign microorganism to helped T lymphocytes.
b. The humeral immunity that characterizes the primary response is coordinated by T lymphocytes.
c. The primary response is quicker and more powerful than the secondary response.
d. Suppressor T cells activate in the secondary response to prevent an overactive immune response.

41. Eosinophils are best described as which of the following?
   a. A type of granulocyte that secretes histamine, which stimulates the inflammatory response.
   b. The most abundant type of white blood cell and they secrete substances that are toxic to pathogens.
   c. A type of granulocyte found under mucous membranes and defends against multicellular parasites.
   d. A type of circulating granulocyte that is aggressive and has high phagocytic activity.

42. Which of the following about epithelial tissue is NOT true?
   a. It does not contain blood vessels.
   b. The cells are more loosely packed than in connective tissue.
   c. It regenerates quickly.
   d. It covers exposed surfaces and lines internal cavities and passageways in the body.

43. How do organisms maintain homeostasis?
   a. They increase their body temperature, blood pH, and fluid balance.
   b. They undergo biochemical processes and absorb energy to increase entropy.
   c. They undergo biochemical processes to maintain the order of their external environment.
   d. They use free energy and matter via biochemical processes to work against entropy.

44. Which of the following correctly lists the four properties that all types of muscle tissue share?
   a. Contractile, excitable, elastic, extensible
   b. Contractile, voluntary, elastic, extensible
   c. Contractile, excitable, voluntary, extensible
   d. Contractile, excitable, elastic, voluntary

45. Which of the following about the autonomic nervous system (ANS) is true?
   a. It controls the reflex arc
   b. It contains motor (efferent) neurons
   c. It contains sensory (afferent) neurons
   d. It contains both parasympathetic nerves and sympathetic nerves

46. What is the function of the sinuses?
   a. To trap the many airborne pathogens
   b. To direct air down the trachea rather than the esophagus
   c. To warm, humidify, and filter air
   d. To sweep away pathogens and direct them toward the top of the trachea

47. Which of the following structures acts like a funnel by delivering the urine from the millions of collecting tubules to the ureters?
   a. The renal pelvis
   b. The renal cortex
   c. The renal medulla
   d. Bowman's capsule

48. Which of the following are functions of the urinary system?
      I. Synthesizing calcitriol and secreting erythropoietin
      II. Regulating the concentrations of sodium, potassium, chloride, calcium, and other ions
      III. Reabsorbing or secreting hydrogen ions and bicarbonate

IV. Detecting reductions in blood volume and pressure

a. I, II, and III
b. II and III
c. II, III, and IV
d. All of the above

49. Which of the following best defines the term *amphoteric*?
a. A substance that conducts electricity due to ionization when dissolved in a solvent
b. A substance that can act as an acid or a base depending on the properties of the solute
c. A substance that, according to the Brønsted-Lowry Acid-Base Theory, is a proton-donor
d. A substance that donates its proton and forms its conjugate base in a neutralization reaction

50. Which of the following touch receptors respond to light touch and slower vibrations?
a. Merkel's discs
b. Pacinian corpuscles
c. Meissner's corpuscles
d. Ruffini endings

Answer Explanations

1. C: The somatic nervous system is the voluntary nervous system, responsible for voluntary movement. It includes nerves that transmit signals from the brain to the muscles of the body. Breathing is controlled by the autonomic nervous system. Thought and fear are complex processes that occur in the brain, which is part of the central nervous system.

2. D: Catabolism is the process of breaking large molecules into smaller molecules to release energy for work. Carbohydrates and fats are catabolized to provide energy for exercise and daily activities. Anabolism synthesizes larger molecules from smaller constituent building blocks. Bioenergetics and metabolism are more general terms involving overall energy production and usage.

3. D: Arteries carry oxygen-rich blood from the heart to the other tissues of the body. Veins carry oxygen-poor blood back to the heart. Intestines carry digested food through the body. Bronchioles are passageways that carry air from the nose and mouth to the lungs.

4. A: The skeletal system consists of the 206 bones that make up the skeleton, as well as the cartilage, ligaments, and other connective tissues that stabilize the bones. The skeletal system provides structural support for the entire body, a framework for the soft tissues and organs to attach to, and acts as a protective barrier for some organs, such as the ribs protecting the heart and lungs, and the vertebrae protecting the spinal cord. The muscular system includes skeletal muscles, cardiac muscle, and the smooth muscles found on the inside of blood vessels. The endocrine system uses ductless glands to produce hormones that help maintain hemostasis, and the reproductive system is responsible for the productions of egg and sperm cells.

5. A: Platelets are the blood components responsible for clotting. There are between 150,000 and 450,000 platelets in healthy blood. When a clot forms, platelets adhere to the injured area of the vessel and promote a molecular cascade that results in adherence of more platelets. Ultimately, the platelet aggregation results in recruitment of a protein called fibrin, which adds structure to the clot. Too many platelets can cause clotting disorders. Not enough leads to bleeding disorders.

6. A: Choice B might be an attractive answer choice, but neutrophils are part of the innate immune system and are not considered part of the primary immune response. The first event that happens in a primary immune response is that macrophages ingest pathogens and display their antigens. Then, they secrete interleukin 1 to recruit helper T cells. Once helper T cells are activated, they secrete interleukin 2 to simulate plasma B and killer T cell production. Only then can plasma B make the pathogen specific antibodies.

7. C: The epididymis stores sperm and is a coiled tube located near the testes. The immature sperm that enters the epididymis from the testes migrates through the 20-foot long epididymis tube in about two weeks, where viable sperm are concentrated at the end. The vas deferens is a tube that transports mature sperm from the epididymis to the urethra. Seminal vesicles are pouches attached that add fructose to the ejaculate to provide energy for sperm. The prostate gland excretes fluid that makes up about a third of semen released during ejaculation. The fluid reduces semen viscosity and contains enzymes that aid in sperm functioning; both effects increase sperm motility and ultimate success.

8. A: All three joint types given are synovial joints, allowing for a fair amount of movement (compared with fibrous and cartilaginous joints). Of the three given, hinge joints, such as the elbow, permit the least motion because they are uniaxial and permit movement in only one plane. Saddle joints and condyloid joints both have reciprocating surfaces that mate with one another and allow a variety of motions in numerous planes, but saddle joints, such as the thumb carpal-metacarpal joint allow more motion than condyloid joints. In saddle joints, two concave surfaces articulate, and in a condyloid joint, such as the wrist, a concave surface articulates with a convex surface, allowing motion in mainly two planes.

9. C: Although the lungs may provide some cushioning for the heart when the body is violently struck, this is not a major function of the respiratory system. Its most notable function is that of gas exchange for oxygen and carbon dioxide, but it also plays a vital role in the regulation of blood pH. The aqueous form of carbon dioxide, carbonic acid, is a major pH buffer of the blood, and the respiratory system directly controls how much carbon dioxide stays and is released from the blood through respiration. The respiratory system also enables vocalization and forms the basis for the mode of speech and language used by most humans.

10. D: All of the above. Synergists are responsible for helping the primary movers or agonists carry out their specific movements. Synergists can help stabilize the point of origin of a muscle or provide extra pull near the insertion; in this sense, they can be considered "fixators." Some synergists can help prevent undesired movement at a joint. For example, for elbow flexion, the brachialis is a synergist to the biceps.

11. D: Bony matrix is an intricate lattice of collagen fibers and mineral salts, particularly calcium and phosphorus. The mineral salts are strong but brittle, and the collagen fibers are weak but flexible, so the combination of the two makes bone resistant to shattering and able to withstand the normal forces applied to it.

12. A: The sternum is medial to the deltoid because it is much closer (typically right on) the midline of the body, while the deltoid is lateral at the shoulder cap. Superficial means that a structure is closer to the body surface and posterior means that it falls behind something else. For example, skin is superficial to bone and the kidneys are posterior to the rectus abdominis.

13. A: When activated, B cells create antibodies against specific antigens. White blood cells are generated in red and yellow bone marrow, not cartilage. Platelets are not a type of white blood cell, and are typically cell fragments produced by megakaryocytes. White blood cells are active throughout nearly all of one's life, and have not been shown to specially activate or deactivate because of life events like puberty or menopause.

14. D: Mechanical digestion is physical digestion of food and tearing it into smaller pieces using force. This occurs in the stomach and mouth. Chemical digestion involves chemically changing the food and breaking it down into small organic compounds that can be utilized by the cell to build molecules. The salivary glands in the mouth secrete amylase that breaks down starch, which begins chemical digestion. The stomach contains enzymes such as pepsinogen/pepsin and gastric lipase, which chemically digest protein and fats, respectively. The small intestine continues to digest protein using the enzymes trypsin and chymotrypsin. It also digests fats with the help of bile from the liver and lipase from the pancreas. These organs act as exocrine glands because they secrete substances through a duct. Carbohydrates are digested in the small intestine with the help of pancreatic amylase, gut bacterial flora and fauna, and brush border enzymes like lactose. Brush border enzymes are contained in the towel-like microvilli in the small intestine that soak up nutrients.

15. B: The structure exclusively found in eukaryotic cells is the nucleus. Animal, plant, fungi, and protist cells are all eukaryotic. DNA is contained within the nucleus of eukaryotic cells, and they also have membrane-bound organelles that perform complex intracellular metabolic activities. Prokaryotic cells (archae and bacteria) do not have a nucleus or other membrane-bound organelles and are less complex than eukaryotic cells.

16. B: The cell structure responsible for cellular storage, digestion and waste removal is the lysosome. Lysosomes are like recycle bins. They are filled with digestive enzymes that facilitate catabolic reactions to regenerate monomers. The Golgi apparatus is designed to tag, package, and ship out proteins destined for other cells or locations. The centrioles typically play a large role only in cell division when they ratchet the chromosomes from the mitotic plate to the poles of the cell. The mitochondria are involved in energy production and are the powerhouses of the cell.

17. D: Veins have valves, but arteries do not. Valves in veins are designed to prevent backflow, since they are the furthest blood vessels from the pumping action of the heart and steadily increase in volume (which decreases the available pressure). Capillaries diffuse nutrients properly because of their thin walls and high surface area and are not particularly dependent on positive pressure.

18. C: The blood leaves the right ventricle through a semi-lunar valve and goes through the pulmonary artery to the lungs. Any increase in pressure in the artery will eventually affect the contractibility of the right ventricle. Blood enters the right atrium from the superior and inferior venae cava veins, and blood leaves the right atrium through the tricuspid valve to the right ventricle. Blood enters the left atrium from the pulmonary veins carrying oxygenated blood from the lungs. Blood flows from the left atrium to the left ventricle through the mitral valve and leaves the left ventricle through a semi-lunar valve to enter the aorta.

19. D: Sodium bicarbonate, a very effective base, has the chief function to increase the pH of the chyme. Chyme leaving the stomach has a very low pH, due to the high amounts of acid that are used to digest and break down food. If this is not neutralized, the walls of the small intestine will be damaged and may form ulcers. Sodium bicarbonate is produced by the pancreas and released in response to pyloric stimulation so that it can neutralize the acid. It has little to no digestive effect.

20. D: A reflex arc is a simple nerve pathway involving a stimulus, a synapse, and a response that is controlled by the spinal cord—not the brain. The knee-jerk reflex is an example of a reflex arc. The stimulus is the hammer touching the tendon, reaching the synapse in the spinal cord by an afferent pathway. The response is the resulting muscle contraction reaching the muscle by an efferent pathway. None of the remaining processes is a simple reflex. Memories are processed and stored in the hippocampus in the limbic system. The visual center is located in the occipital lobe, while auditory processing occurs in the temporal lobe. The sympathetic and parasympathetic divisions of the autonomic nervous system control heart and blood pressure.

21. C: Nociceptor are a type of touch receptor that detect pain. They relay this information to the CNS via afferent nerves. Efferent nerves then carry motor signals back, which may cause muscle contractions that move the body part away from the source of the pain. For example, if a person places their hand on a hot stove, nociceptors detect the painful burning sensation. Then they relay this information up to the brain, where a response signal travels down the efferent nerves that control the muscles of the shoulder, arm, forearm, and/or hand to retract the hand from the hot stove. Choice *A* is incorrect because that is mainly the function of Ruffini endings. Choice *B* describes the function of several mechanoreceptors such as Meissner's corpuscles, Pacinian corpuscles, and Merkel's discs. Lastly, thermoreceptors detect changes in pressure, so Choice *D* is incorrect.

22. B: Ligaments connect bone to bone. Tendons connect muscle to bone. Both are made of dense, fibrous connective tissue (primary Type 1 collagen) to give strength. However, tendons are more organized, especially in the long axis direction like muscle fibers themselves, and they have more collagen. This arrangement makes more sense because muscles have specific orientations of their fibers, so they contract in somewhat predictable directions. Ligaments are less organized and more of a woven pattern because bone connections are not as organized as bundles or muscle fibers, so ligaments must have strength in multiple directions to protect against injury.

23. A: The three primary body planes are coronal, sagittal, and transverse. The coronal or frontal plane, named for the plane in which a corona or halo might appear in old paintings, divides the body vertically into front and back sections. The sagittal plane, named for the path an arrow might take when shot at the body, divides the body vertically into right and left sections. The transverse plane divides the body horizontally into upper or superior and lower or inferior sections. There is no medial plane, per se. The anatomical direction medial simply references a location close or closer to the center of the body than another location.

24. A: The strength of an aqueous acid or base depends on the extent to which its electrolytes dissociate in water. The more complete the ionization, the stronger the solution it forms. Choice *C* is incorrect because it describes the strength of an acid, while Choice *D* is incorrect because it describes the strength of a base.

25. A: The forebrain contains the cerebrum, the thalamus, the hypothalamus, and the limbic system. The limbic system is chiefly responsible for the perception of emotions through the amygdala, while the cerebrum interprets sensory input and generates movement. Specifically, the occipital lobe receives visual input, and the primary motor cortex in the frontal lobe is the controller of voluntary movement. The hindbrain, specifically the medulla oblongata and brain stem, control and regulate blood pressure and heart rate.

26. A: It is most likely asthma. Any of the answer choices listed can cause heavy breathing. A blood clot in the lung (*B*) could cause this, but this would be very uncommon for a child. Choices *C* and *D* can both be ruled out because the question mentions that it occurs even when the patient is relaxing. Hyperventilation is usually caused by a panic attack or some sort of physical activity. Asthma often develops during childhood. It would stand to reason then that the child may have not yet been diagnosed. While asthma attacks can be caused by exercise they can also occur when a person is not exerting themselves.

27. C: 2:9:10:4. These are the coefficients that follow the law of conservation of matter. The coefficient times the subscript of each element should be the same on both sides of the equation.

28. B: Substances with higher amounts of hydrogen ions will have lower pHs, while substances with higher amounts of hydroxide ions will have higher pHs. Choice *A* is incorrect because it is possible to have an extremely strong acid with a pH less than 1, as long as its molarity of hydrogen ions is greater than 1. Choice *C* is false because a weak base is determined by having a pH lower than some value, not higher. Substances with pHs greater than 2 include anything from neutral water to extremely caustic lye. Choice *D* is false because a solution with a pH of 2 has ten times fewer hydrogen ions than a solution of pH 1.

29. A: Salts are formed from compounds that use ionic bonds. Disulfide bridges are special bonds in protein synthesis which hold the protein in their secondary and tertiary structures. Covalent bonds are strong bonds formed through the sharing of electrons between atoms and are typically found in organic molecules like carbohydrates and lipids. London dispersion forces are fleeting, momentary bonds which occur between atoms that have instantaneous dipoles but quickly disintegrate.

30. C: As in the last question, covalent bonds are special because they share electrons between multiple atoms. Most covalent bonds are formed between the elements H, F, N, O, S, and C, while hydrogen bonds are formed nearly exclusively between H and either O, N, or F of other molecules. Covalent bonds may inadvertently form dipoles, but this does not necessarily happen. With similarly electronegative atoms like carbon and hydrogen, dipoles do not form, for instance. Crystal solids are typically formed by substances with ionic bonds like the salts sodium iodide and potassium chloride.

31. A: The upper back has one of the highest densities of sweat glands of any area on the body. While palms, arms, and feet are often thought of as sweaty areas, they have relatively low amounts of sweat glands compared to other parts of the body. Remember that one of the purposes of sweat is thermoregulation, or controlling the temperature of the body. Regulating the temperature of one's core is more important than adjusting the temperature of one's extremities.

32. C: Nephrons are responsible for filtering blood. When functioning properly they allow blood cells and nutrients to go back into the bloodstream while sending waste to the bladder. However, nephrons can fail at doing this, particularly when blood flood to the kidneys is limited. The medulla (also called the renal medulla) (*A*) and the renal cortex (*D*) are both parts of the kidney but are not specifically responsible for filtering blood. The medulla is in the inner part of the kidney and contains the nephrons. The renal cortex is the outer part of the kidney. The heart (*B*) is responsible for pumping blood throughout the body rather than filtering it.

33. B: Ossification is the process by which cartilage, a soft, flexible substance is replaced by bone throughout the body. All humans regardless of age have cartilage, but cartilage in some areas goes away to make way for bones.

34. C: Water's polarity lends it to be extremely cohesive and adhesive; this cohesion keeps its atoms very close together. Because of this, it takes a large amount of energy to melt and boil its solid and liquid forms. Phospholipid bilayers are made of nonpolar lipids and water, a polar liquid, cannot easily flow through it. Cell membranes use proteins called aquaporins to solve this issue and let water flow in and out. Fish breathe by capturing dissolved oxygen through their gills. Water can self-ionize, wherein it decomposes into a hydrogen ion (H^+) and a hydroxide ion (OH^-), but it cannot self-hydrolyze.

35. A: The neutrons and protons make up the nucleus of the atom. The nucleus is positively charged due to the presence of the protons. The negatively charged electrons are attracted to the positively charged nucleus by the electrostatic or Coulomb force; however, the electrons are not contained in the nucleus. The positively charged protons create the positive charge in the nucleus, and the neutrons are electrically neutral, so they have no effect. Radioactivity does not directly have a bearing on the charge of the nucleus.

36. D: The seminiferous tubules are responsible for sperm production. Had *testicles* been an answer choice, it would also have been correct since it houses the seminiferous tubules. The prostate gland (*A*) secretes enzymes that help nourish sperm after creation. The seminal vesicles (*B*) secrete some of the components of semen. The scrotum (*C)* is the pouch holding the testicles.

37. C: The parasympathetic nervous system is related to calm, peaceful times without stress that require no immediate decisions. It relaxes the fight-or-flight response, slows heart rate to a comfortable pace, and decreases bronchiole dilation to a normal size. The sympathetic nervous system, on the other hand, is in charge of the fight-or-flight response and works to increase blood pressure and oxygen absorption.

38. C: The parathryroid gland impacts calcium levels by secreting parathyroid hormone (PTH). Osteotoid gland is not a real gland. The pineal gland regulates sleep by secreting melatonin, and the thymus gland focuses on immunity. *Thyroid* would also be a correct answer choice as it influences the levels of circulating calcium.

39. B: The endocrine system's organs are glands which are spread throughout the body. The endocrine system itself does not connect the organs or transport the hormones they secrete. Rather, the various glands secrete the hormone into the bloodstream and lets the cardiovascular system pump it throughout the body. The other three body systems each include a network throughout the body:

- Cardiovascular system: veins and arteries
- Immune system: lymphatic vessels (it does also use the circulatory system)
- Nervous system: nerve networks

40. D: In the secondary immune response, suppressor T lymphocytes are activated to negate the potential risk of damage to healthy cells, brought on by an unchecked, overactive immune response. Choice *A* is incorrect because the activity is characteristic of the primary response, not the secondary response. Choice *B* is incorrect because humeral immunity is mediated by antibodies produced by B, not T, lymphocytes. Choice *C* is wrong because the secondary response is faster than the primary response because the primary response entails the time-consuming process of macrophage activation.

41. C: Eosinophils, like neutrophils, basophils, and mast cells, are a type of leukocyte in a class called granulocytes. They are found underneath mucous membranes in the body and they primarily secrete destructive enzymes and defend against multicellular parasites like worms. Choice *A* describes basophils and mast cells, and Choice *B* and *D* describe neutrophils. Unlike neutrophils which are aggressive phagocytic cells, eosinophils have low phagocytic activity.

42. B: Epithelial tissue includes epithelia, which are the layers of cells that cover exposed surfaces of the body and line internal cavities, passageways, and glands. Epithelia do not contain blood vessels. This tissue can regenerate quickly to replace damaged and dead cells. The cells in epithelial tissue are tightly packed together, while in connective tissue they are sparser and scattered throughout the extracellular matrix.

43. D: The natural tendency in the universe is to increase entropy and disorder, so organisms undergo biochemical processes and use free energy and matter to stabilize their internal environment against this changing external environment in a process called homeostasis. Organisms strive to maintain physiologic factors such as body temperature, blood pH, and fluid balance in equilibrium or around a set point. Choice *A* is incorrect because homeostasis in organisms involves trying to maintain internal conditions around their set point, which doesn't always involve increasing body temperature, blood pH, and fluid balance; the organism may need to work to decrease these values. Choice *B* is incorrect because maintaining homeostasis expends energy; it does not absorb it. Choice *C* is incorrect because homeostasis involves maintaining order in the organism's internal, not external, environment.

44. A: All three types of muscle tissue (skeletal, cardiac, and smooth) share four important properties: they are contractile (they can shorten and pull on connective tissue); they are excitable (they respond to stimuli); they are elastic (they rebound to their original length after a contraction); and they are extensible (they can be stretched repeatedly but maintain the ability to contract). While skeletal muscle is under voluntary control, both cardiac and smooth muscle are involuntary.

45. D: The autonomic nervous system (ANS), a division of the PNS, controls involuntary functions such as breathing, heart rate, blood pressure, digestion, and body temperature via the antagonistic parasympathetic and sympathetic nerves. Choices *A, B,* and *C* are incorrect because they describe characteristics of the somatic nervous system, which is the other division of the PNS. It controls skeletal muscles via afferent and efferent neurons. Afferent neurons carry sensory messages from skeletal muscles, skin, or sensory organs to the CNS, while Efferent neurons relay motor messages from the CNS to skeletal muscles, skin, or sensory organs. While skeletal muscle movement is under voluntary control, the somatic nervous system also plays a role in the involuntary reflex arc.

46. C: The sinuses function to warm, filter, and humidify air that is inhaled. Choice *A* is incorrect because mucus traps airborne pathogens. Choice *B* is incorrect because the epiglottis is the structure in the pharynx that covers the trachea during swallowing to prevent food from entering it. Lastly, Choice *D,* sweeping away pathogens and directing them toward the top of the trachea, is the function of cilia. Respiratory structures, such as the nasal passages and trachea, are lined with mucus and cilia.

47. A: The renal pelvis acts like a funnel by delivering the urine from the millions of the collecting tubules to the ureters. It is the most central part of the kidney. The renal cortex is the outer layer of the kidney, while the renal medulla is the inner layer. The renal medulla contains the functional units of the kidneys—nephrons—which function to filter the blood. Choice *D,* Bowman's capsule, is the name for the structure that covers the glomeruli.

48. D: The urinary system has many functions, the primary of which is removing waste products and balancing water and electrolyte concentrations in the blood. It also plays a key role in regulating ion concentrations, such as sodium, potassium, chloride, and calcium, in the filtrate. The urinary system helps maintain blood pH by reabsorbing or secreting hydrogen ions and bicarbonate as necessary. Certain kidney cells can detect reductions in blood volume and pressure and then can secrete renin to activate a hormone that causes increased reabsorption of sodium ions and water. This serves to raise blood volume and pressure. Kidney cells secrete erythropoietin under hypoxic conditions to stimulate red blood cell production. They also synthesize calcitriol, a hormone derivative of vitamin D3, which aids in calcium ion absorption by the intestinal epithelium.

49. B: An amphoteric substance can act as an acid or a base depending on the properties of the solute. Water is a common example and because of its amphoteric property, it serves as a universal solvent. Choice *A* is incorrect because it describes electrolytes. Choices *C* and *D* are incorrect because they both describe an acid.

50. C: The body has a variety of touch receptors that detect tactile information, which gets carried via general somatic afferents and general visceral afferents to the CNS. Meissner's corpuscles respond to light touch and slower vibrations. Choice *A* is incorrect because Merkel's discs respond to sustained pressure. Choice *B* is incorrect because Pacinian corpuscles detect rapid vibration in the skin and fascia, and Choice *D* is incorrect because Ruffini endings detect deep touch and tension in the skin and fascia.

Dear HAPS Test Taker,

We would like to start by thanking you for purchasing this study guide for your HAPS exam. We hope that we exceeded your expectations.

Our goal in creating this study guide was to cover all of the topics that you will see on the test. We also strove to make our practice questions as similar as possible to what you will encounter on test day. With that being said, if you found something that you feel was not up to your standards, please send us an email and let us know.

We would also like to let you know about other books in our catalog that may interest you.

GRE

This can be found on Amazon: amazon.com/dp/162845900X

MCAT

amazon.com/dp/1628456779

GMAT

amazon.com/dp/1637758847

We have study guides in a wide variety of fields. If the one you are looking for isn't listed above, then try searching for it on Amazon or send us an email.

Thanks Again and Happy Testing!
Product Development Team
info@studyguideteam.com

Printed in the USA
CPSIA information can be obtained
at www.ICGtesting.com
LVHW021541170823
755429LV00013B/465

9 781637 751848